Embedded Control System Design

Alexandru Forrai

Embedded Control System Design

A Model Based Approach

Springer

Author
Alexandru Forrai
Mitsubishi Elevator Europe
Veenendaal
The Netherlands

ISBN 978-3-642-28594-3 e-ISBN 978-3-642-28595-0
DOI 10.1007/978-3-642-28595-0
Springer Heidelberg New York Dordrecht London

Library of Congress Control Number: 2012933081

Printed on acid-free paper

Springer is part of Springer Science+Business Media (www.springer.com)

Preface

This book emerged from lecture notes prepared for tutorials and seminars mainly targeting engineers working in the industry, involved in embedded control system design. One of the goals is to present a unified theory for control system design that captures the essential issues and can be applied for a wide range of practical problems. The control design techniques presented in the book are all model based. There are several approximations involved in obtaining a mathematical model of the plant. A simple transfer function for the plant is often desired: the reason behind is that for complicated plant models controller design is difficult.

Whenever is possible a model for control shall be derived from measured input/output data, using for example system identification experiments.

The approximations and simplifications in the plant dynamics lead to a nominal plant model. However, if the controller designed for the nominal model does not take into account the approximation errors (called plant uncertainties) then the feedback system may become unstable when this controller is used for the real plant. In order to avoid this situation, we should determine a bound on the uncertainty and use this information when we design the controller.

The control problem can be solved in:

- a classical control framework, in which the controller design is performed for the nominal plant according to the performance specifications. Later on stability analysis and performance evaluation is done taking into account the plant uncertainty, too.
- robust control framework, in which given a plant and an uncertainty bound characterizing the set of all possible plants, we want to derive a controller, which stabilizes the feedback system for the set of all possible plants and the closed-loop system achieves the desired performance specifications.

All these aspects are emphasized in the book.

Next, the derived controller shall be implemented in real-time, sophisticated control algorithms are difficult to implement, are numerically fragile and are not really appreciated in the industry.

Real-time implementation of the control algorithm leads the control engineer into the world of embedded software. Therefore, we considered important to describe good software development practices according to the generic standard IEC61508. Furthermore, we dedicated an essential part of the book to safety issues, especially development of embedded safety loops according to the generic standard IEC61508.

The lecture notes are addressed to practicing control engineers, embedded software engineers and students in engineering who are interested in control system design and implementation. It is assumed that the reader followed an undergraduate course in signals and systems, and is familiar with frequency-domain methods of analyzing feedback control systems, such as Bode plots, Nyquist criterion, state-space representation, etc. To keep the development simple, the considered systems are: single-input/single-output systems, linear-time-invariant or linear-parameter-varying systems.

Acknowledgements

I am grateful for many individuals. First of all, I would like to express my gratitude to my former professors from Technical University of Cluj-Napoca, Romania: Prof. M. Ivan who introduced me many years ago into the theory of linear spaces and reviewed carefully the chapter dedicated to system identification and made valuable remarks, Prof. Emeritus T. Colosi, who introduced me into control theory and who reviewed the chapters dealing with classical control design. I am indebted to Prof. V. Trifa my Ph.D. supervisor and Prof. Emeritus I.A. Viorel with whom I worked together in good cooperation for several years at Technical University of Cluj-Napoca, Romania. Furthermore, I would like to take this opportunity to thank for the permanent support of Prof. R. Munteanu, rector of Technical University of Cluj-Napoca, Romania. Next, I would like to express my high appreciation to my former professors and colleagues from Utsunomiya University, Japan: Prof. Emeritus Prof. K. Kamiyama, Prof. S. Adachi (currently Keio University, Japan), Prof. H. Funato and Prof. S. Hashimoto (currently Gunma University, Japan) for their guidance, help and support during my stay in Japan. In addition, I would like to thank to Prof. Emeritus K. Kamiyama for his comments and remarks after reviewing the chapter dedicated to robust control. Furthermore, I would like to thank to Prof. D. Simon from Department of Electrical and Computer Engineering Cleveland State University, U.S.A. and Mr. G. Ristroph from IJK Controls, U.S.A. for their comments and remarks after reviewing the chapters dedicated to classical control design. Special thanks to Prof. J.A. Abraham from the University of Texas at Austin, U.S.A. for providing his paper related to memory testing.

I am also deeply indebted to Prof. I. Nagy and Prof. P. Korondi from Budapest University of Technology and Economics, Hungary, for their unconditional help and support in the past years. I highly appreciate the support and friendship of my former colleague Dr. G. Dozsa, currently with ARM Ltd. Cambridge, U.K. who reviewed carefully some of the book chapters. I would like to thank to Mr. T. Ueda, department manager at Mitsubishi Elevator Europe, The Netherlands, for his kindness and support in the past years. His support, especially in very difficult times of my private life is highly appreciated. In addition, I would like to thank to all my colleagues and former colleagues for the opportunity to work together and learn from them in the past twenty years. Many thanks to my friends from The Netherlands: Ms. Rodica Galan, Mr. Richard Roest, Mr. Bert van der Poel and his family for their support and advice in the past years. In would like to thank to the editors of this book: Dr. Thomas Ditzinger and Mr. Holger Schpe for their understanding and support and for the good work of the cover designer Ms.Jessica Wengrzik. Last but not least, I am highly indebted to family - to my wife Yuko-san and my son Keiji-kun -for their support, understanding and love. Furthermore, I would like to thank to my father, who never stopped loving and helping me. I would dedicate this book to the memory of my mother, who was all her life a cheerful, loving and supporting mother and from whom I learned to love to read and listen to music.

February, 2012 Alexandru Forrai
Utrecht, The Netherlands

Contents

1 Embedded Systems – Basic Concepts 1
1.1 What Is an Embedded System? 1
1.2 The Main Architecture of Embedded Control Systems 2
1.2.1 The Electric Power Level 3
1.2.2 The Signal Processing Level 5
1.3 Communication Networks in Embedded Systems 9
1.3.1 The Main Features of a CAN Communication Network 9
1.3.2 CAN Message Frames 12
1.3.3 Error Detection and Signaling 13
1.3.4 CAN Controller Modes 14
1.3.5 CAN Implementations 15
1.4 Multi-tasking Embedded Control Systems 15
1.5 Planning Embedded System Development 17

2 Introduction into Embedded Control System Design 21
2.1 Requirements for Control System Design 21
2.1.1 Safety Requirements 21
2.1.2 Identification of the System to Be Controlled 23
2.1.3 Control Device Specification 26
2.1.4 Design 27
2.1.5 Installation and Maintenance 33
2.2 Mathematical Models for Control 34
2.2.1 Models from Science 34
2.2.2 Models from Experimental Data 34
2.2.3 Linearization of Nonlinear Models 36
2.3 Control System's Characteristics 40
2.3.1 Disturbance Attenuation 42
2.3.2 Tracking 42
2.3.3 Sensitivity to Parameter Variations 43
2.3.4 Control System's Limitation 44

2.3.5 Stability and Relative Stability 48
2.4 Performance Specifications for Linear Systems 51

3 **System Identification and Model-Order Reduction** 55
3.1 Model Building and Model Structures 56
3.1.1 Model Structures 58
3.2 Input Signal Design for System Identification Experiments ... 61
3.2.1 Requirements Imposed on the Input Signal 63
3.2.2 Input Signal Design 65
3.3 Model Validation in Time and Frequency Domain 69
3.4 Model-Order Reduction Methods 75
3.4.1 Model-Order Reduction 75
3.5 Nominal Plant and Plant Uncertainties 81
3.5.1 Multiplicative Uncertainty Model 84
3.5.2 Additive Uncertainty Model 85
3.6 Practical Examples – System Identification 88
3.6.1 Brushless d.c. Drive's Identification 88
3.6.2 Identification of a Fuel Cell 92
3.6.3 Short Review and Remarks 96

4 **Classical Controller Design - Part I** 99
4.1 Controller Design Based on Pole-Zero Cancellation 99
4.1.1 The Influence of Controller Zero 100
4.2 Controller Design for Deadbeat Response 103
4.3 Controller Design Using the Root Locus Technique 106
4.3.1 Phase-Lead Controller Design Using the Root Locus .. 109
4.3.2 Phase-Lag Controller Design Using the Root Locus ... 112
4.4 PID Controller Design 115
4.4.1 Ziegler-Nichols Tuning Formula 115
4.4.2 Chien-Hrones-Reswick Tuning Formula 117
4.5 The Coefficient Diagram Method 117
4.6 Validation of the Control System 122
4.6.1 Representative Sample and Sample Size 122
4.6.2 Monte Carlo Simulation 124

5 **Classical Controller Design - Part II** 127
5.1 Controller Design for Systems with Time Delays 127
5.1.1 Systems with Time Delays – Smith Predictor 129
5.1.2 The Coefficient Diagram Method (CDM) for Systems with Time Delays 131
5.1.3 Handling Jitter in Networked Control System 135
5.2 Controller Design for Disturbance Rejection 138
5.2.1 Notch Filters 138
5.3 Disturbance Observers 140
5.3.1 Electromagnetic Armature Movement Detection 143

5.4 Two-Degree-of-Freedom Control Systems (2DOF) 144
5.5 Control System Design Verification and Validation 150

6 Fundamentals of Robust Control 153
6.1 Review of Norms for Signals and Systems 153
6.1.1 Norms for Signals 154
6.1.2 Norms for Systems 154
6.2 Internal Stability 156
6.2.1 Youla Parametrization 159
6.2.2 Youla Parametrization for Systems with Time Delays 161
6.3 Unstructured Plant Uncertainties 162
6.3.1 Unstructured Additive Plant Uncertainty 163
6.3.2 Unstructured Multiplicative Plant Uncertainty 163
6.4 Robust Stability 165
6.4.1 Robust Stability for Different Uncertainty Models 166
6.5 Robust Performance 168
6.5.1 Performance and Robustness Bounds 170
6.6 Design for Robust Performance 172
6.6.1 Performance Weighting Function Design for Tracking Control 174
6.6.2 Performance Weighting Function Design for Disturbance Rejection 175
6.7 Robust Controller Synthesis Problem 176
6.7.1 Mixed-Sensitivity Optimization for Tracking Problem 177
6.7.2 Mixed-Sensitivity Optimization for Disturbance Rejection 183

7 Robust Controller Design 187
7.1 Controller Design Using Youla Parametrization 187
7.2 Controller Design Using Robust Control Toolbox 190
7.2.1 Robust Controller Design for Brushless D.C. Drive 193
7.2.2 Robust Controller Design for Active Vibration Suppression 196
7.3 Controller Design with Constraint on the Control Signal 206
7.4 Robust Gain-Scheduled Control 208
7.4.1 Theoretical Background 209
7.4.2 Stability Analysis 211
7.4.3 Robust Gain-Scheduled Control Design – Example 212
7.5 Control Algorithm Implementation in Real-Time 216

8 Embedded Safety Loop Development 219
8.1 Risk Assessment and Safety Levels 219
8.1.1 Safety Function and Safety Loop 221
8.2 Classification of Faults 223
8.2.1 Fault Categories and Failure Rates 223
8.2.2 Safe Failure Fraction, Diagnostic Coverage and Hardware Fault-Tolerance 225
8.3 Calculation of Probability of Failure on Demand 228
8.3.1 PFH Calculation for Single-Channel System 228
8.3.2 PFH Calculation for Dual-Channel System 229
8.3.3 Operating Demand of the Safety Function 236
8.4 Software Testing and Validation 237
8.4.1 Static and Dynamic Software Testing Methods 237
8.4.2 Software Development Stage Related Testing Methods 238
8.4.3 Black-Box and White Box Testing 239
8.4.4 Software Testing Metrics – Code Coverage 239
8.4.5 Hardware in the Loop Tests 242
8.5 Memory Testing 246
8.5.1 RAM Fault Models 247
8.5.2 Testing for Stuck-at Faults 249
8.5.3 Testing for Coupling Faults 251

References 257

Chapter 1

Embedded Systems – Basic Concepts

'A journey of a thousand miles begins with a single step. It does not matter how slowly you go, so long as you do not stop.' — Confucius (551-479 BC)

Summary:

This chapter introduces the reader into embedded systems in general and embedded control systems in particular.

The following topics are discussed and briefly explained:

- what is an embedded system and what is an embedded control system?
- the architecture of an embedded control system
- embedded communication networks, with main focus on Control Area Network (CAN)
- multi-tasking embedded systems
- project planning.

1.1 What Is an Embedded System?

What is an embedded system?

The embedded system could be defined as a processor based system designed to perform a few dedicated functions, often in real-time. It can be found in complex devices such as mobile phones, MP3 players, traffic lights, home appliances, automobiles, etc.

Embedded systems are usually built around microcontroller units (abbreviated as MCUs) or digital signal processors (abbreviated as DSPs), depending on the application.

Since embedded systems are designed for specific tasks and often mass-produced they are optimized in size, production costs, performance and reliability. Some embedded systems are simple, have only a few inputs and outputs, others are more complex and are interconnected via a communication network with other embedded systems.

Developing embedded systems requires knowledge in different fields. Usually the development is done in a multidisciplinary team involving hardware designers, software designers, control engineers, quality engineers, etc.

Since the field of embedded systems is very wide we are going to focus only on embedded systems applied in control applications.

The next and obvious question is: what is an embedded control system? The embedded control system is an embedded system dedicated to control in real-time a system or a sub-system.

In this book our main focus is on closed-loop control system design and implementation in real-time. We are not going to discuss embedded hardware or software design issues, but we will provide necessary background information - useful for control engineers.

The targeted audience is mainly practicing control engineers, software engineers involved in real-time implementation of control algorithms, undergraduate and graduate students interested in embedded control systems. However, the book is not an introductory book into control system theory, it is assumed that the reader completed such a course and is familiar with control theory and the related mathematical background.

Last but not least, some book chapter might contain useful information for project managers involved in embedded system development.

1.2 The Main Architecture of Embedded Control Systems

In the aim to illustrate the main architecture of an embedded control system, let us consider a simplified vehicle control system shown in Fig. 1.1.

We remark the hierarchical control structure of the vehicle control system. On highest hierarchical level is the driver control, which provides reference value(s) for the embedded control system. Current, position, speed and other sensor signals are measured and are fed into the embedded control system. All these functions are situated at the signal processing level.

Based on the reference value(s) and on the measured signals the control system computes the new reference value for the power converter, which supplies the electrical machine (electric power level). The electrical machine via the transmission system drives the host vehicle (mechanical level).

Next, very briefly we will discuss the signal processing level and the electric power level. The mechanical level is out of our scopes, therefore is not discussed here.

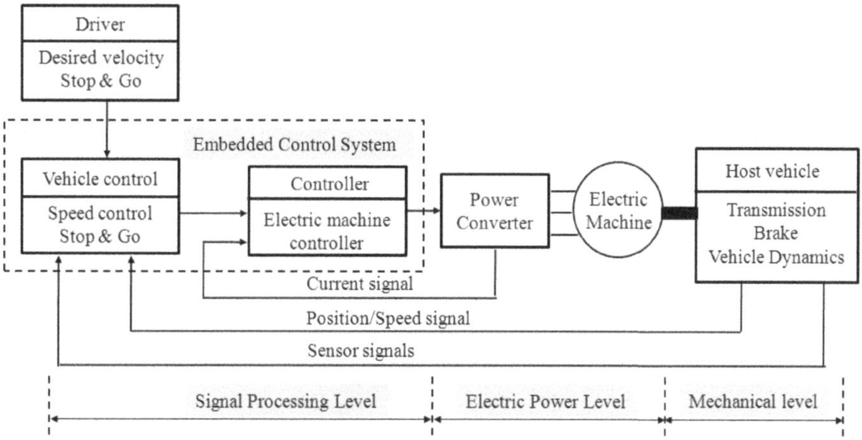

Fig. 1.1 Vehicle control system.

1.2.1 The Electric Power Level

For simplicity, let us consider that the electric machine is a direct current (d.c.) machine, supplied via a so-called H-bridge shown in Fig. 1.2.

By changing the applied voltage on the d.c. machine, we can change its speed: the speed ω is almost proportional with the applied medium voltage V_{med}.

The applied voltage (its medium value) is changed via pulse-width-modulation (PWM) by switching ON and OFF the constant supply d.c. voltage V_{DC} as shown in Fig. 1.3.

The duty factor is defined as:

$$Duty[\%] = \frac{T_{ON}}{T_{ON} + T_{OFF}} 100[\%] \tag{1.1}$$

The medium voltage is:

$$V_{med} = \frac{T_{ON}}{T_{ON} + T_{OFF}} V_{DC} \tag{1.2}$$

In the H-bridge, if the switching devices denoted by T_1 and T_4 are switched then the machine will rotate in one direction and if the switching devices denoted by T_2 and T_3 are switched then the machine will rotate in the opposite direction.

Typical current waveforms are also shown in Fig. 1.3, where I_{min}, I_{med} and I_{max} corresponds to minimum, medium and maximum current values.

We remark that the motor current I_{med} depends on the mechanical load and does not depend on the applied voltage. The electromagnetic torque T_e of the d.c. machine is proportional with the motor current I_{med}.

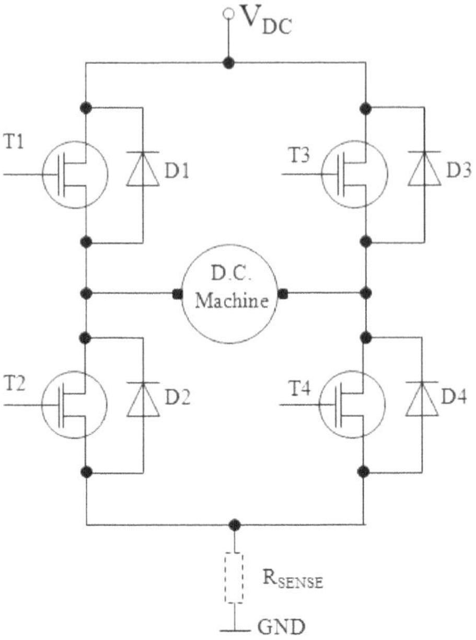

Fig. 1.2 D.C. machine connected to an H-bridge.

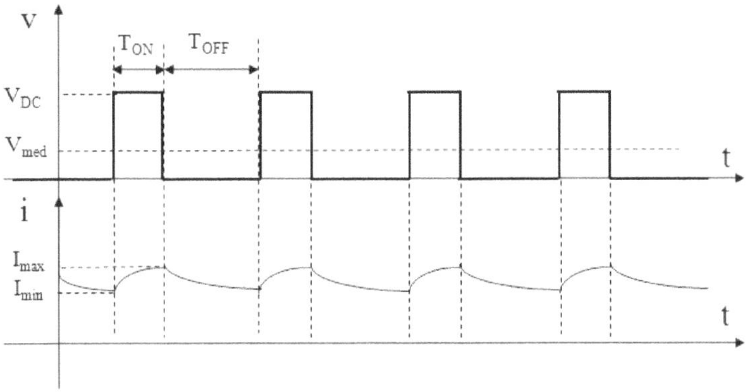

Fig. 1.3 Pulse width modulation - voltage and current waveforms.

The machine can operate in motor mode or generator mode (see in Fig. 1.4). Using the H-bridge a four quadrant operation is possible: quadrant I and III corresponds to motor mode and II and IV corresponds to generator mode. Therefore, if the embedded control system provides the right switching sequence with the correct duty cycle is possible to control the machine speed and direction of movement.

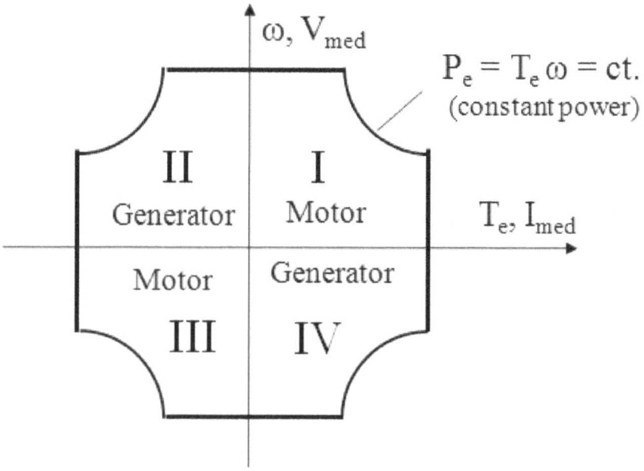

Fig. 1.4 Operation modes of the d.c. machine.

1.2.2 The Signal Processing Level

The embedded control system is built around a central processing unit (CPU) as shown in Fig. 1.5. When the CPU and and different kind of input/output ports are integrated on the same chip we are talking about a microcontroller unit (MCU).

In this particular example, we have input/output ports such as: analog inputs, digital inputs, pulse width modulated (PWM) outputs and a communication port for diagnostics purposes.

Usually port-pin configuration (i.e. digital port 1 pin 0 is assigned to be digital input) is realized via so-called special function registers (SFRs), which are MCU specific and detailed information is provided by the vendor, so it is not discussed here.

Other essential components of the embedded system are: the power supply voltage monitor, the watchdog timer and the communication port which are explained further on.

Power Supply Voltage Monitor

The power supply voltage monitoring is part of most embedded systems. A very simple power supply monitor is shown in Fig. 1.6, when V_{CC} falls below the reset threshold V_{TH}, the output RESET goes low and holds the microcontroller in reset.

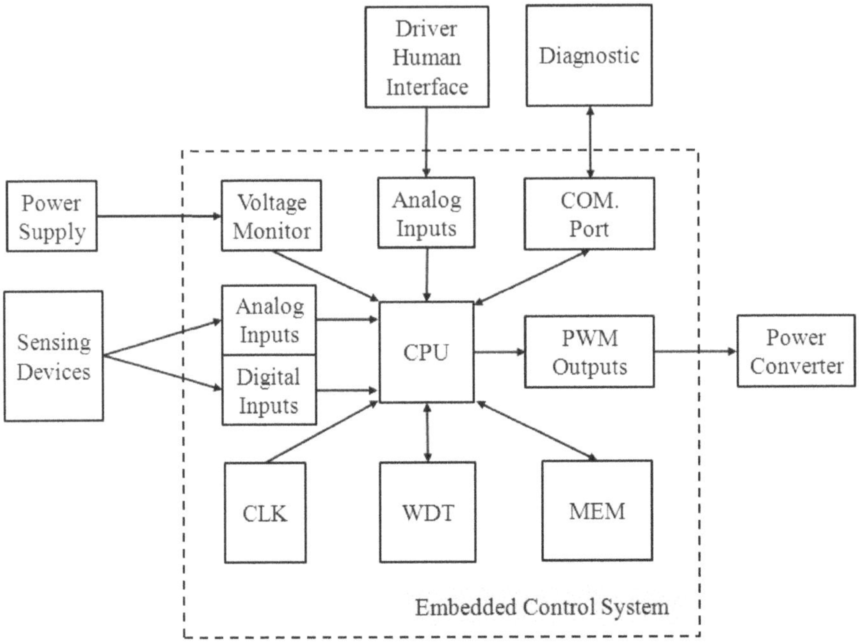

Fig. 1.5 Embedded system architecture.

In applications is also required that on power-up the RESET output is held low for example 150 ms after V_{CC} rises above the threshold.

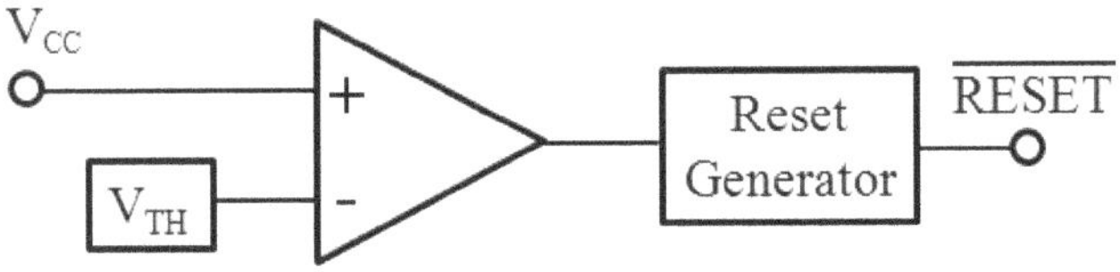

Fig. 1.6 Simple power supply monitor.

A more sophisticated circuit is the LTC 2911 (by Linear Technology) is a low power, high accuracy triple supply monitor with power-fail comparator. The V1 pin monitors a 3.3V supply. The V2 pin monitors a 5V, 2.5V, 1.8V, 1.2V or adjustable supply. A third adjustable input (ADJ) has a nominal 0.5V threshold allowing the resistive divider to configure its threshold [89] as shown in Fig. 1.7.

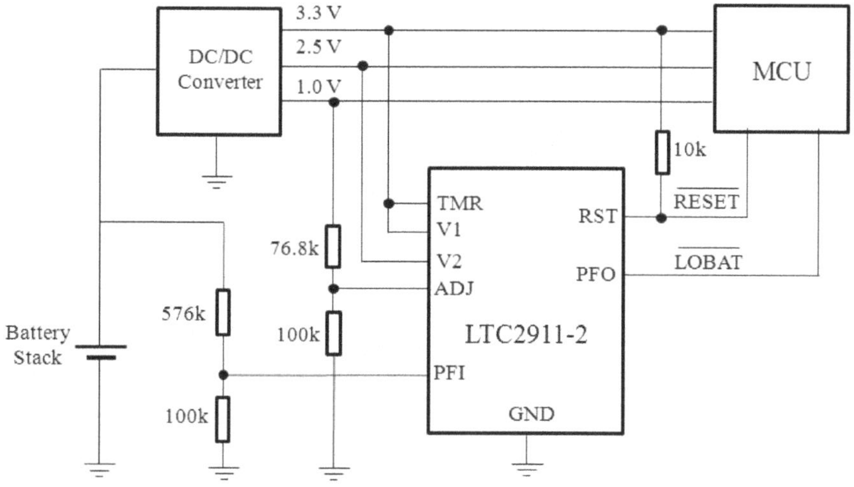

Fig. 1.7 Voltage monitor.

Furthermore, a spare comparator can be configured to provide early warning of a low voltage condition. It causes the power failure output (PFO) to pull low when the voltage of the power failure input (PFI) falls below 0.5V.

Watchdog Timer

Next the main features of the watchdog timer are briefly discussed. Most embedded systems must be self-reliant, that means if the software hangs-on or crashes, the system shall reboot automatically. This supervision is achieved by an external watchdog timer.

Basically, a watchdog timer is based on a counter that counts down from some initial value to zero. The microcontroller shall service the watchdog within a so-called "time-window" otherwise the watchdog will reset the microcontroller.

A typical window-watchdog operating circuit is shown in Fig. 1.8 [90].

It is recommended to reset the watchdog at different points in the program rather than servicing the watchdog input periodically by using the pre-scaler of a microcontroller.

The TPS3813 watchdog timer from Texas Instruments has implemented window-watchdog settings, depending how the WDT and WDR pins are connected. There are four different timings available with these settings as mentioned in Table 1.1.

In addition to the implemented window-watchdog settings, the upper boundary of the watchdog timer can be set by an external capacitor C_{ext} connected between the WDT pin and GND.

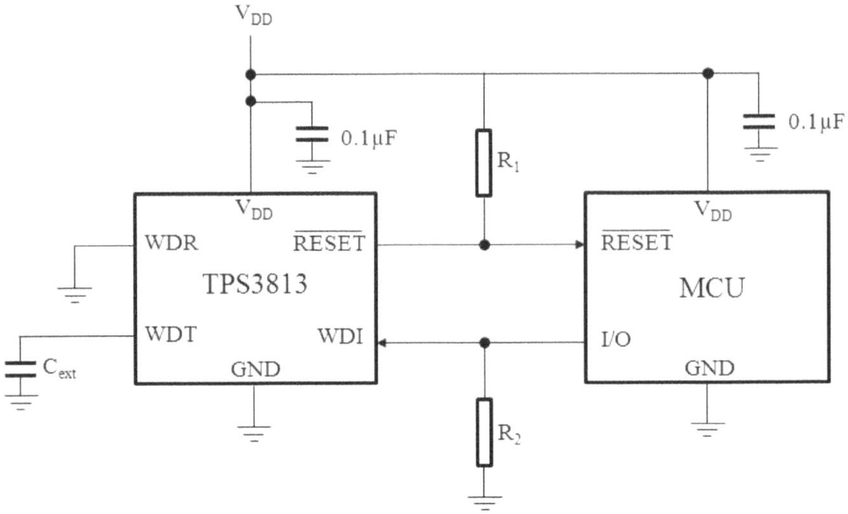

Fig. 1.8 Window-watchdog circuit.

Table 1.1 Watchdog window frame configuration

Selected Operation Mode		Window frame	Lower window frame
$WDT = 0V$	WDR = 0 V	Max = 0.3 s	Max = 9.46 ms
		Typ = 0.25 s	Typ = 7.86 ms
		Min = 0.2 s	Min = 6.27 ms
	$WDR = V_{DD}$	Max = 0.3 s	Max = 2.43 ms
		Typ = 0.25 s	Typ = 2 ms
		Min = 0.2 s	Min = 1.58 ms
$WDT = V_{DD}$	WDR = 0 V	Max = 3 s	Max = 93.8 ms
		Typ = 2.5 s	Typ = 78.2 ms
		Min = 2 s	Min = 62.5 ms
	$WDR = V_{DD}$	Max = 3 s	Max = 23.5 ms
		Typ = 2.5 s	Typ = 19.6 ms
		Min = 2 s	Min = 15.6 ms

For an embedded application, the important boundaries are the $T_{boundary\_max}$ and T_{window_min}. Within these boundaries the watchdog shall be serviced (re-triggered). In the shaded area (see Fig. 1.9) as well as between the boundary tolerance (dashed area) it cannot be predicted that the watchdog will release a reset.

The next topic, which we would briefly discuss are communication networks in embedded systems. Since communication networks play a central role in embedded control systems, the next section is fully dedicated to this topic, especially to CAN networks.

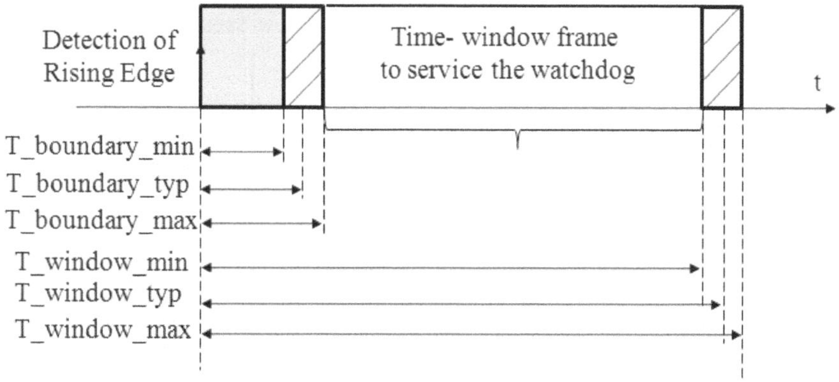

Fig. 1.9 Upper and lower boundaries for a watchdog timer.

1.3 Communication Networks in Embedded Systems

In practice, often embedded control systems are interconnected via a communication network. There are different communication protocols, which are used in different industry sectors [86].

CAN (Control Area Network) is a very popular communication protocol and is frequently encountered during embedded system design. Therefore, we will briefly introduce it here.

In Fig. 1.10 we show a simplified CAN communication network in a vehicle.

CAN is a serial communication network, originally designed for the automotive industry, which became popular in industrial automation as well as in other industrial fields. It is a two-wire, half duplex, high-speed network system and is well-suited for high speed applications using short messages. Its robustness, reliability and the availability of low price CAN controllers made it to spread widely. It offers high-speed communication rate up to 1 Mbit/sec and thus allows real-time control.

1.3.1 The Main Features of a CAN Communication Network

The main features of a CAN communication network are:

- multi master network
- communication rate up to 1 Mbit/sec
- network length is typically 40 m at 1 Mbit/sec and up to 10 km at 5 kbit/sec
- high reliability with extensive error checking
- the maximum latency of high priority message is less than 120 microseconds at 1 Mbit/sec transmission rate.

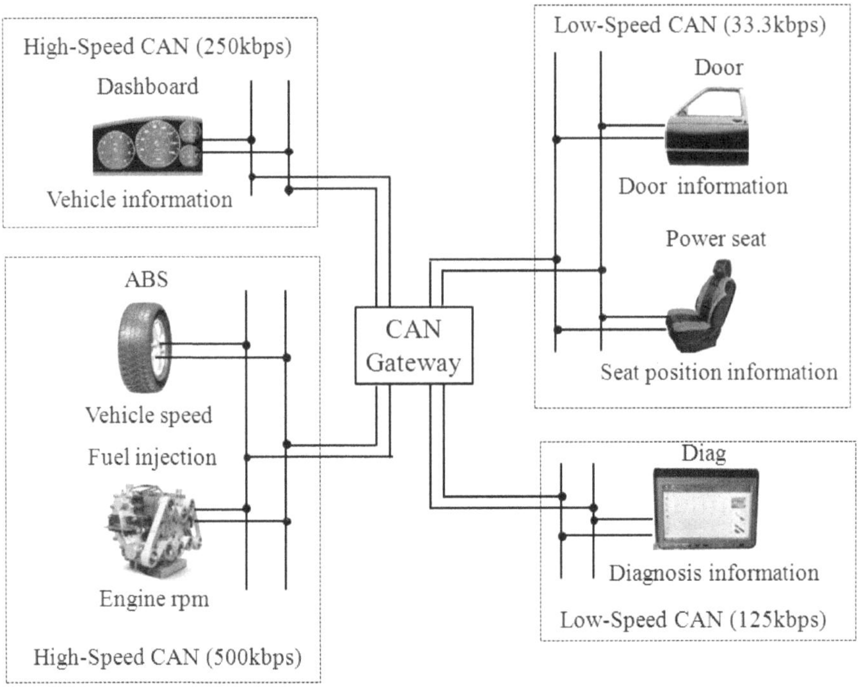

Fig. 1.10 CAN communication network in a vehicle.

The CAN nodes do not have a specific address, instead the CAN messages have unique identifiers, which uniquely identify the message and determines the message priority. Lower message identifiers (MID) are associated with higher message priority and vice versa.

Before sending a message the CAN node checks if the bus is busy. If in the same time two or more messages are sent then message collision happens, which is solved by arbitration.

The way in which message collision is handled is that each node transmits its message identifier (MID) and listens on the bus while transmitting. If it is a collision, the message with lower MID (higher priority) will hold the signal down (zero bit is said to be dominant) and the messages with higher MID (lower priority) will stop transmitting.

There are two CAN specifications in use:

- Standard CAN (version 2.0A) specification, which is known as standard CAN with 11 bit message identifier, operated at a maximum frequency of 125 kbit/sec ISO11519.
- Extended CAN (version 2.0B) specification, known as full CAN having extended CAN frame with 29 bit message identifier, which can be used up to 1 Mbit/sec ISO11898.

The main differences are: different formats of the message frame (message identifier length) and differences in the physical layer (transmission speed).

The CAN network may be implemented over a number of physical media, but the most common media is a twisted pair, 5V differential signal, which will allow operations in high noise environment and will work even if one of the wires is open circuit. Widely used CAN transceiver chips are Philips 82C251 and TJA1040.

The voltage levels for a CAN network, which follows the ISO 11898 (CAN High Speed) standard are shown in Table 1.2.

Table 1.2 CAN voltage levels according to ISO 11898

Signal	Recessive state			Dominant state		
	Min.	Nominal	Max	Min	Nominal	Max
CAN-High	2.0 V	2.5 V	3.0 V	2.75 V	3.5 V	4.5 V
CAN-Low	2.0 V	2.5 V	3.0 V	0.5 V	1.5 V	2.25 V

When running full CAN (ISO11898-2) at higher speeds, it is required to terminate the CAN bus at both ends with 120Ω resistors (as shown in in Fig. 1.11). The termination resistors prevent reflections and unload the open collector transistors. Note that for the recessive state, nominal voltage for the two wires is the same. This decreases the power drawn from the nodes through the termination resistors.

The maximum bus length for a CAN network depends on the bit rate used. Below in Table 1.3 for different bus lengths the corresponding maximum bit rates are listed.

Table 1.3 CAN bus length function on the transmission rate

Bus length (in metres)	Maximum bit rate (bit/s)
40	1 Mbit/s
100	500 kbit/s
200	250 kpit/s
500	125 kbit/s
6000	10 kbit/s

1.3.2 CAN Message Frames

CAN is using two types of message frames: normal frames and remote frames.

The normal frames are used to carry data and contain the following fields (see also Fig. 1.11):

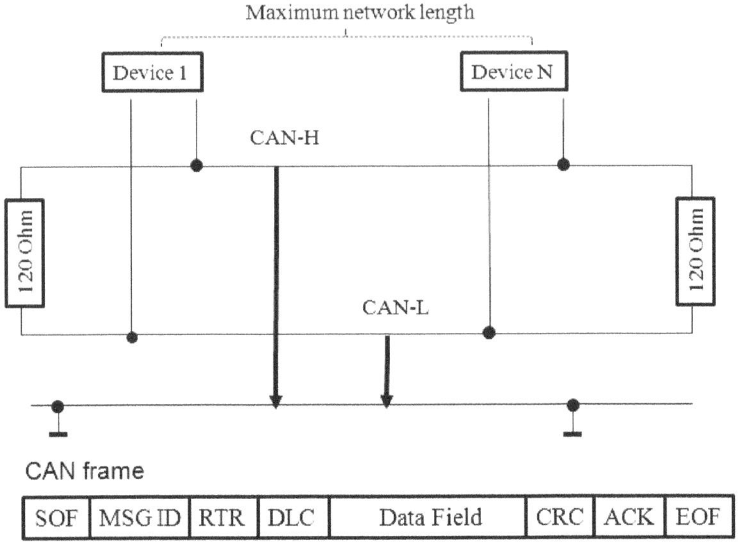

Fig. 1.11 CAN network and CAN frame structure.

Normal Frame

- Start of frame (SOF)
- Message Identifier (MID) either 11 or 29 bits long depending on the chosen mode
- Remote Transmission Request (RTR) = 0, see also remote frames for nonzero
- Data Length Control (DLC) the number of data bytes to follow (0-8)
- Data Field (DATA)
- CRC field containing a 15 bit cyclic redundancy check code
- Acknowledge field (ACK) an empty slot, which will be filled by the receiving node
- End of Frame (EOF)

Remote frames are used to request a particular message to be put on the network. The fields of a remote frame are:

Remote Frames

- Start of frame (SOF)
- Message identifier (MID) 11 bits or 29 bits depending on the chosen mode
- Remote Transmission Request (RTR) = 1
- Data Length Control (DLC) the number of expected data bytes to be returned (0-8)
- CRC field containing a 15 bit cyclic redundancy check code
- Acknowledge field (ACK) an empty slot, which will be filled by the receiving node
- End of Frame (EOF)

1.3.3 Error Detection and Signaling

CAN is a very reliable system with multiple error checking. The built in error detection of the controllers together with the error signaling make sure that the information is correct and consistent. Faulty nodes will go to error modes where they do not disturb the traffic on the bus.

Error detection is handled automatically by the CAN controller. The detected errors are:

- bit errors
- message errors.

The bit errors are divided into two categories:

- bit stuffing error - normally a transmitting node inserts a high bit after five consecutive low bits (and a low after five consecutive high). This is called bit stuffing. A receiving node that detects more than five consecutive bits (high or low) will see a bit stuffing violation.
- bit error - a transmitting node always reads back the message as it is sending. If it detects a different bit value on the bus than it sent, and the bit is not part of the arbitration field or in the acknowledgment field, and error is detected.

The message errors are divided as:

- checksum error - each receiving node checks CAN messages for checksum errors.
- frame error - there are certain predefined bit values that must be transmitted at certain points within any CAN Message Frame. If a receiver detects an invalid bit in one of these positions a Form Error (also known as a Format Error) will be flagged.
- acknowledgment error - if a transmitter determines that a message has not been acknowledged then an ACK Error is flagged.

1.3.4 CAN Controller Modes

The CAN controller can be in one of three modes (states):

- error active - the normal operating mode for a controller. Messages can be received and transmitted. On detecting an error an active error flag is sent.
- error passive - a mode entered when the controller has frequent problems transmitting or receiving messages. Messages can be received and transmitted. On detecting an error while receiving, a passive error flag is sent.
- bus off - entered if the controller has serious problems with transmitting messages. No messages can be received or transmitted until the CAN controller is reset by the host microcontroller or processor.

When an error is detected by a node it sends an error flag on the bus. This prevents any other node from accepting the message and ensures consistency of data throughout the network.

The active error flag consists of six low bits, and is used if the node transmitting the error frame is in active error state. As low is dominant all other nodes will detect bit stuffing violation and send their own error flag. After this, nodes that want to transmit (including the one sending the interrupted message) will start to do so. As usual, the node whose message identifier has the highest priority will win arbitration and send its message.

If the CAN controller is in error passive mode the error frame will consist of six passive (high) bits. Since the error flag only consists of passive bits, the bus is not affected. If no other node detected an error, the message will be sent uninterrupted. This ensures that a node having problems with receiving cannot block the bus.

All of this advanced error handling is done automatically by the CAN controller, without any need for the host microcontroller to do anything. This is one of the big advantages of CAN.

The mode of the CAN controller is controlled by two error counters - the transmit error counter (TX_count) and the receive error counter (RX_count). The following rules apply:

- the CAN controller is in error active mode if TX_count $<=$ 127 AND RX_count $<=$ 127.
- passive mode is used if (TX_count $>$ 127 OR RX_count $>$ 127) AND TX_count $<=$ 255.
- bus off is entered if TX_count $>$ 255.

Once the CAN controller has entered bus off state, it must be reset by the host microcontroller or processor in order to be able to continue operation. In addition, this is only allowed after the reception of 128 occurrences of 11 consecutive recessive bits. How the transmit error counter (TX_count) and receive error counter (RX_count) are incremented for details, the reader shall refer to the standard [87].

1.3.5 CAN Implementations

There are different CAN implementations used in different industrial sectors, we will mention only two of them:

DeviceNet was developed for use in industrial process control and is based on the standard Full CAN - ISO 11898-2.

DeviceNet rigorously defines the physical interconnect, has a more restrictive transceiver specification, uses 11 bit identifiers only, allows 125, 250 and 500kbit/sec operation only and regulates the message content in order to more easily support interoperability of different manufacturers units.

CANopen was designed with control applications in mind, it is a software standard based on the standard Full CAN - ISO 11898-2. It limits the number of nodes to 127 and allocates them IDs.

Profiles are specified for each type of device by CAN in Automation (CiA) organization in order to simplify the use of similar units (e.g. motor drives) from different manufacturers. Later on, we will also refer to CANopen Safety, which is a safety communication protocol based on CANopen.

1.4 Multi-tasking Embedded Control Systems

Till now, we have seen the main architecture of an embedded control system and we have discussed some of its essential components.

Embedded control systems represent an important sub-class of real-time systems. In case of real-time systems not only the correctness of the computation but also the time at which the result is available is important. The temporal behavior of the real-time system relies on the availability of the resources [55].

A possibility to handle this issue is to divide the control algorithm into tasks and prioritize them. In this way the tasks will have to compete (based on defined priorities) for the resources of the embedded processor.

Such a multi-tasking environment is illustrated in Fig. 1.12.

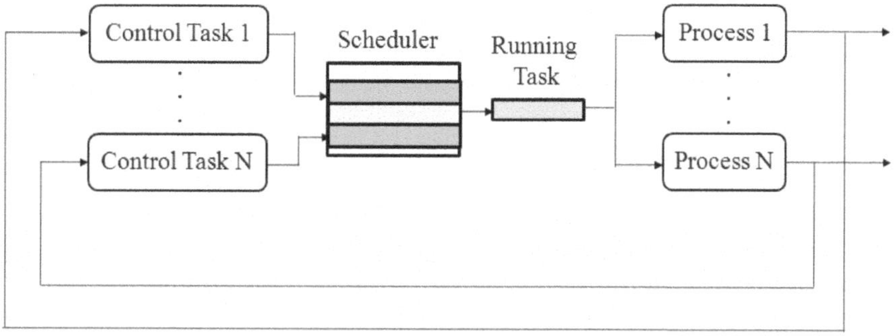

Fig. 1.12 Multitasking embedded control system.

For complex embedded control projects is recommend to use a real-time operating system (RTOS), which provides task scheduling, task dispatching and inter-task communication.

Traditionally, the controller is designed by control engineers and the implementation is the responsibility of system (software) engineers. It might happen that - worst case scenario - the control engineers pay no attention on how the designed control algorithm will be implemented, while the software engineers do not have an idea about the temporal requirements of the algorithms.

However, having a RTOS or so some kind of task scheduler will impose a framework on the software architecture and will create a much clear temporal behavior of the control algorithm. It also, offers more flexibility for the designers during the development process to manage properly the available resources.

The entity deciding which task shall be started is called scheduler.

The scheduler decides - based on task priority - which is the next task, which is going to run. In OSEK (RTOS widely used in the automotive industry) the value 0 is defined as the lowest priority and higher numbers define higher priorities [88]. A very simple task model (basic task in OSEK) is illustrated in Fig. 1.13.

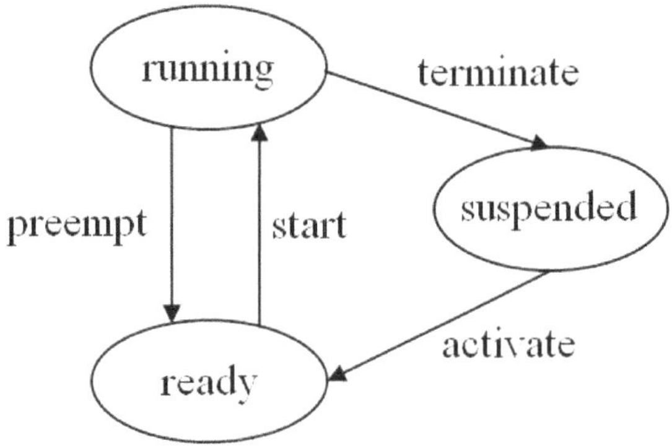

Fig. 1.13 Task model - the states of a task.

The main states of a task are:

- running state the task can be or is executed by the embedded processor. Only one task can be in this state at any point in time.
- ready state - all conditions necessary to transit into the running state exist. The task waits for allocation of the processor. The scheduler decides which ready task to be executed next.
- suspended state the task is passive and can be activated.

A full preemptive (priority based) scheduling means that a task having higher priority being in ready state, will preempt the running lower priority task. The preempted task (lower priority task) context is saved. After the higher priority task is completed the lower priority task can run from the location where it was preempted (illustrated in Fig. 1.14).

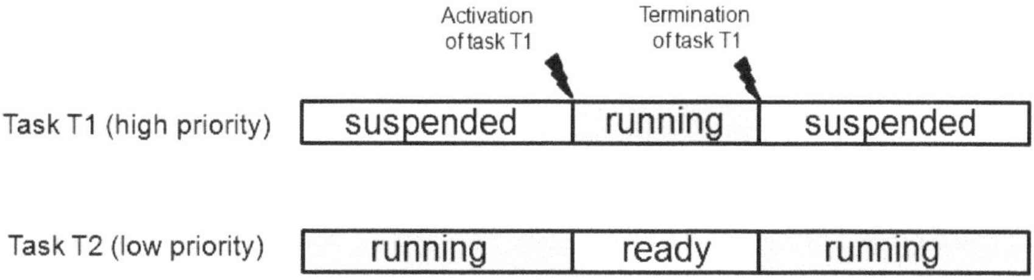

Fig. 1.14 Full preemptive task scheduling.

1.5 Planning Embedded System Development

It was mentioned that embedded system development is a multi-disciplinary field and requires knowledge from different areas and interactions with people having different technical background.

Embedded systems are developed usually in teams, so proper communication and good project management is essential for project success.

As the magnitude of the project grows, more and more people are involved in the project and the natural expectation is that workload is shared and the project can be finished in time.

However, as clearly mentioned in [34] and sustained by practical experience is: the growth in communication channels between team members sinks productivity on large projects. If in the project there are only two engineers then they need only one communication channel. In case of n engineers there will be:

$$\frac{n(n-1)}{2} \tag{1.3}$$

communication channels.

For example, in a small team having 6 members, there will be 15 communication channels. If the team has 10 members the number or communication channels will be 45. The number of communication channels is increasing heavily (almost square-law) if the number of team members is increased, which will sink the productivity.

The next big challenge described here is scheduling and planning the project.

Several companies face the situation that they have to develop a new embedded system according to the customer needs - but they are not very familiar with the application field. Without clear understanding of the application and requirements it is not wise to make a project plan [34].

A good and recognized approach is the Wideband Delphi (WD) method, which is shown in Fig. 1.15 and described below. The WD approach recognizes that the judgment of experts can be surprisingly accurate but individuals often suffer of bias, and groups might exhibit follow leader behavior [34].

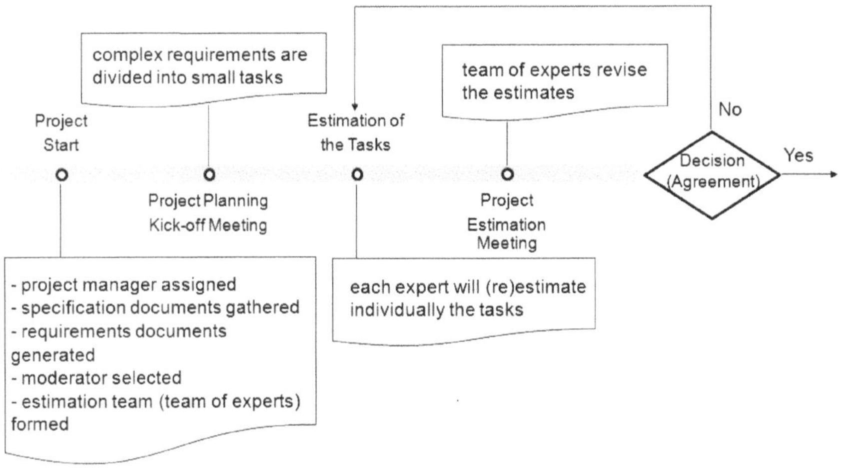

Fig. 1.15 Project planning using the Wideband Delphi method.

According to the Wideband Delphi method a project manager selects a moderator and an estimation team with three to five members. The team is formed by people familiar with the application field and experienced developers.

The process starts by gathering specifications documents and generating requirements documents. The Delphi process - iterative process - consists of at least three meetings run by the moderator.

The first meeting is the kickoff meeting, during which the estimation team creates a work breakdown structure and discusses assumptions. The work breakdown structure is developed by starting with the end objective and successively subdividing it into manageable components in terms of size, duration, and responsibility (e.g., systems, subsystems, components, tasks, subtasks, and work packages) which include all steps necessary to achieve the objective.

Basically, the team creates a high level design in order to divide complex requirements into small tasks, each of which gets estimated via the WD process. After the meeting, each team member estimates the effort for each task.

The second meeting is the estimation session, in which the team revise the estimates as a group and achieves consensus.

The cycle repeats until either no estimator wants to change his or her estimate, the estimators agree that the range is acceptable or two hours have elapsed.

The task estimation, performed in estimation rounds and expressed in manhours [mh] is illustrated in Fig. 1.16.

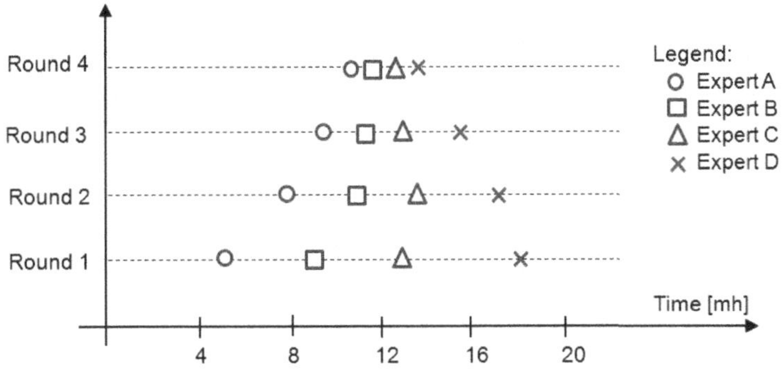

Fig. 1.16 Task estimation in rounds.

After agreement, the project manager summarizes the results and reviews them with the team, at which point they are ready to be used as the basis for planning the project.

Chapter 2

Introduction into Embedded Control System Design

'Understanding what is not possible is as important as understanding what is possible.' — G.C. Goodwin - Bode Lecture, 1999.

Summary:

This chapter introduces the reader into embedded control system design The following items are discussed:

- requirements and safety requirements for control systems
- mathematical models for control
- the main features of control systems
- performance specification - in time and frequency domain - of control system.

2.1 Requirements for Control System Design

2.1.1 Safety Requirements

The first and most important item to consider - during control system design - is safety. It is required to minimize the risk at least to the accepted (tolerable) risk as shown in Fig. 2.1.

Therefore, all applicable local and national codes that regulate the safety during installation and operation of control system shall be followed.

Due to the importance of the topic, a separate chapter will be dedicated to the development of embedded safety loops (see Fig. 2.2). To achieve high safety and integrity level, the safety loop is independent from the control loop and is implemented on a separate hardware.

During design, at least the following items shall be kept in mind and measures shall be taken:

- minimize/eliminate the risk of fire and electrical shock to personnel - provide a quick way to disconnect the power to the equipment or process by using fuses, circuit breakers or other circuit interrupting means

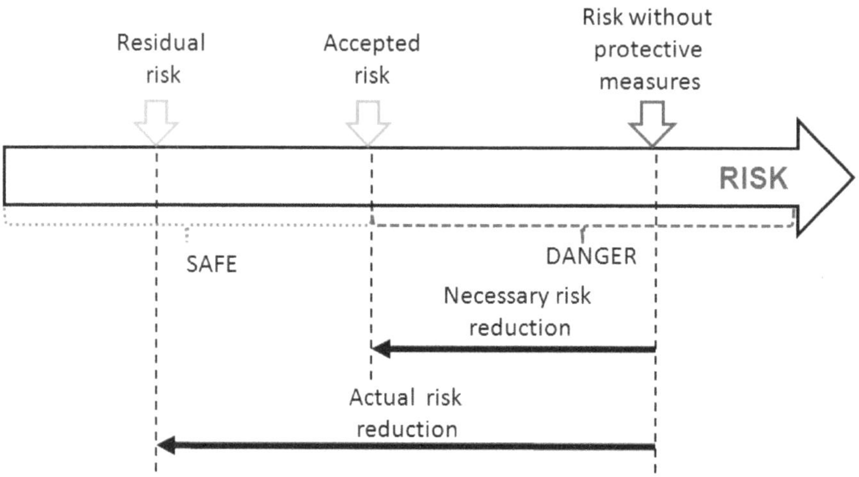

Fig. 2.1 Risk assessment and risk reduction.

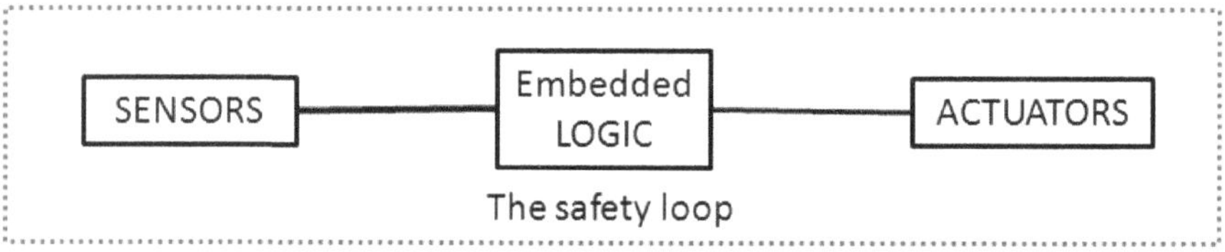

Fig. 2.2 The safety loop.

- safeguard operators and maintenance personnel from unexpected startup (energization) or the release of hazardous energy during service or maintenance activity
- provide emergency stop button - a quick way to disconnect manually the power to machinery, equipment or process
- provide safe shut-down sequence if fault is detected - during shut-down the system shall not create risk of personal injury or equipment damage, if such a situation exists the equipment shall be stopped before personnel come in contact
- provide proper grounding to prevent electrical shock
- provide proper control power distribution - if a fault occurs, only the faulty circuit shall be interrupted (de-energized) the other circuits shall operate
- avoid unexpected operations - reduce risks, avoid personal injury and damage by using devices such as relays and limit switches
- use independent safety loops - from safety viewpoint the designer shall not fully rely on the embedded control system.

Last but not least, is the responsibility of the control engineer to ensure safe shut-down of the system - without causing personnel injury or equipment damage - if the control loop opens (i.e. feedback signal is lost).

The next important item in the control system design is identifying the system, which is going to be controlled.

2.1.2 Identification of the System to Be Controlled

Control engineering is a multi-disciplinary field, a good understanding of basic electricity and safety, hardware and software, mechanics, hydraulics and pneumatics is often required. Furthermore, a deep knowledge of the system, which is going to be controlled, is essential in many cases.

Complexity of the system can be measured by the number of inputs and outputs. A system having only a few inputs and outputs is easier to control than a system with many inputs and outputs.

From control engineering viewpoint is very important to understand, what are the system parameters and specifications, what are the inputs and outputs, how the inputs affect the outputs, are there nonlinearities in the system?

Let us illustrate this by considering a simple system, a conveyor belt, which is powered by a 3-phase motor shown in Fig. 2.3.

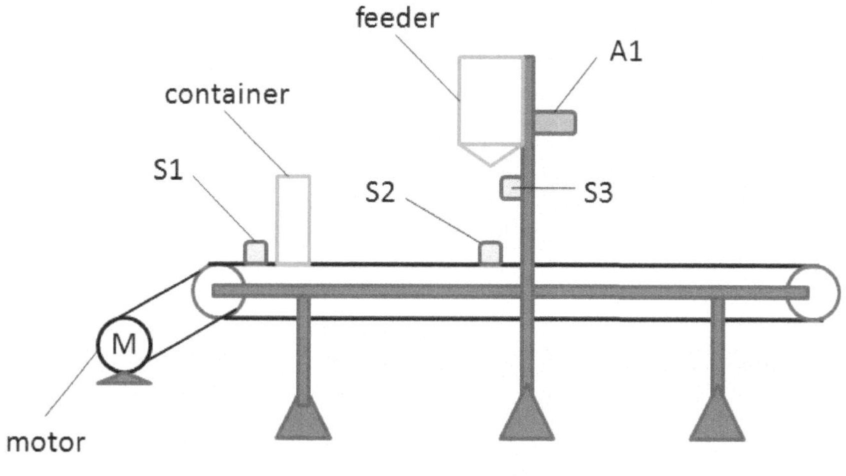

Fig. 2.3 The conveyor belt system.

On the conveyor belt, the container is sensed at two different positions (left and middle) by the sensors S1 and S2. In the middle position (sensed by sensor S2) the container is filled, actuator A1 opens the valve and when the container is full (sensed by sensor S3) closes the valve.

The first step in identification is information gathering about the system or process.

Step 1: Information Gathering

We would like to gather information about the equipment parameters and specifications, for example:

- what is the available voltage?
- what is the available power rating?
- what is the required speed for the conveyor belt?
- what is the range of mechanical load on the conveyor belt?
- what are the mechanical dimensions?
- what are the minimum mounting clearances?
- what is the operating temperature range?
- what is the relative humidity?
- what is the duty cycle?
- who will use the control system?
- how the system will be used?

The next step is identifying the control architecture: how control devices interact and which is the required control sequence.

Step 2: Identifying the Control Architecture and Control Sequence

To automate the conveyor belt we will need different control devices, providing input and output signals for the processing unit. These are listed below, mentioning in parenthesis the signal names:

- a manually operated start and stop button (START/STOP)
- a 3-phase motor sized for the conveyor belt
- a electrically controlled motor starter, including short circuit and overload protection, sized for the 3-phase motor (Motor START/STOP)
- an actuator to control the feeder's valve, including short circuit protection (A1 OPEN/CLOSE)
- sensors that will indicate the container position on the belt and when the container filled (S1,S2 and S3)
- some sort of processing unit (embedded hardware and software).

For simplicity we noted the signals coming from the sensors with S1, S2, S3. The system inputs and outputs are presented in Fig. 2.4.

Let us assume that the container position and content is sensed by the proximity sensors and the process is automated using only ON/OFF control. From functional viewpoint we can describe the system by a finite state machine. The state diagram is shown in Fig. 2.5, where:

- Start Feed = START and (not STOP) and S2 and (not S3)
- Stop Feed = STOP or S3

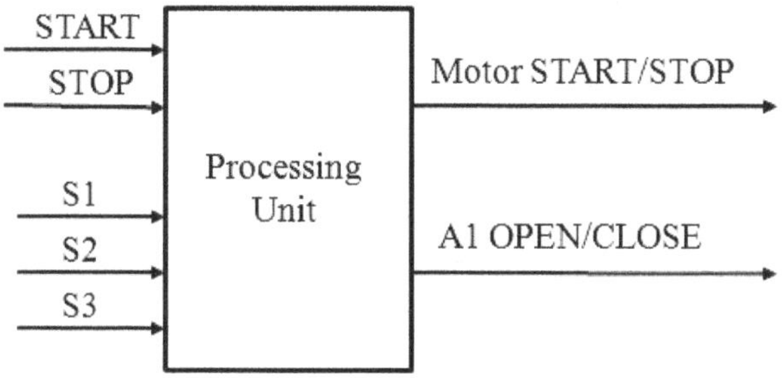

Fig. 2.4 The inputs and outputs of the system.

- Start Move = Start Move 1 or Start Move 2, where
  Start Move 1 = START and (not STOP) and S1 and (not S2)
  Start Move 2 = START and (not STOP) and S2 and S3
- Stop Move = STOP or S2.

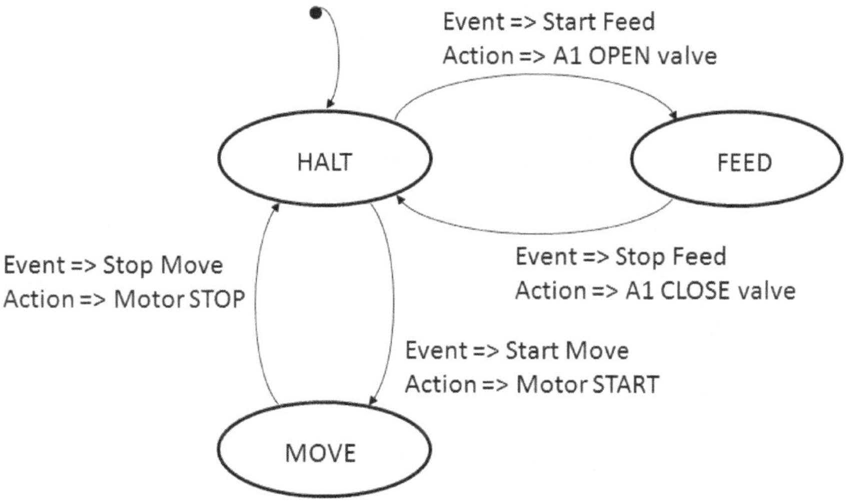

Fig. 2.5 State diagram associated to the conveyor belt.

Furthermore, let us assume that an additional sensor (S4) is added at the end of the conveyor belt. This sensor will be used to count the filled containers. It is left for the reader to extend the control architecture and re-draw the state diagram.

The next step is the control device specification.

2.1.3 Control Device Specification

There are three categories of devices, which have to be specified.

Input Devices

Typical input devices are: sensors, limit switches, operator interface, etc. The specifications shall contain:

- general specifications such as:
  performance level (i.e. PL e, according to EN ISO 13849-1)
  sensing distances (i.e. up to 5mm)
  intended time of usage (i.e. 10 years, according to EN ISO 62061)
- electrical specifications such as:
  supply voltage (i.e. 12V to 30V DC)
  current consumption (i.e. max. 100 mA)
  power consumption (i.e. max. 1.2 Watt)
- mechanical specifications such as:
  weight (i.e. 500g)
  dimensions (i.e. diameter 8mm, length 25mm)
  max. shaft loading (i.e. axial 10 N, radial 50N)
  shock (i.e. less than 100 g half-sine, 6ms, according to EN 60068-2-7)
- environmental specifications such as:
  operating temperature (i.e. -30 +70 Celsius degrees)
  storage temperature (i.e. -30 +70 Celsius degrees)
  humidity (i.e. 98%)
  protection class (IP65, according to EN60529).

Output Devices

The output devices are used to control actions. Typical output devices are relays, motor starters, valves, numerical displays, user displays, etc. In case of inductive devices such as electromagnetic valves is important to specify:

- what is the operating voltage?
- what is the maximum current?
- what is the operating temperature range?
- what are the mounting dimensions?
- what is the duty cycle?

The duty cycle of the actuator A1 - in the conveyor belt example - shall be in accordance with the duty cycle of the process.

A solenoid valve designed for 25% duty cycle shall not be operated in continuous mode with on time 10 seconds and off time 5 seconds. The short off time will not allow the solenoid to cool down and finally it will fail. Solenoid valves and motors are inductive loads and create voltage spikes, which can damage output devices, therefore diodes shall be used for surge suppression.

Processing Unit

The processing unit might be built around a microcontroller unit (MCU) or digital signal processor (DSP) or can be a programmable logic controller (PLC).

Since embedded systems are optimized for size and costs (see Fig. 2.6), it is essential how to choose the right processing unit for a given application. In the example above (conveyor belt application) a simple 8-bit MCU or a simple PLC will be sufficient.

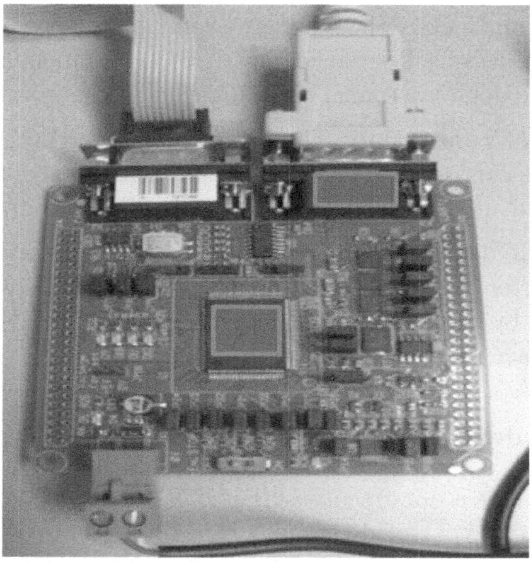

Fig. 2.6 Embedded system with microcontroller unit.

The next important item is design and contains several design phases and tasks.

2.1.4 Design

Design contains several design phases such as:

- electric design
- embedded hardware development
- control algorithm design
- embedded software development.

Electrical Design

The electrical design also contains several steps such as:

Electric schematic design: The electric schematic shows how the control power is distributed to various devices, how devices are connected to ground and how are protected by relays and fuses. Proper grounding of devices is essential in control system design, providing safety and functionality. Furthermore, the design shall fulfill electromagnetic compatibility (EMC) requirements - this is also an essential part of the design - since electromagnetic interference may cause faulty operation or device failure.

Panel layout design: Once the schematic is finalized a panel layout drawing is made at scale. This specifies the layout of various components, so mounting and wiring can be done in a panel or subpanel.

Bill of materials generation: The bill of materials (BOM) will list each component of the control system, quantity of each component and "marks" that allow us to identify the component on the schematic drawing.

Wiring diagram design: Wiring diagram is mainly used for installation by the electrician for routing and terminating the wiring between the various devices. Terminal blocks shall be used (sometimes color coded) to show signals that enter and leave the control panel.

Embedded Hardware Development

Since hardware design is out of scope of the book, let assume that the hardware platform on which the control algorithm will be implemented is available. If the hardware is not available - in the first stages - a so-called evaluation board can be used for software development.

Control Algorithm Development

The process of control algorithm development can be represented by a V-model, shown in Fig. 2.7, which is adopted from the well-know V-model used in software development.

In this book chapters are dedicated to modeling (how to build models for control) and controller design. Furthermore, controller implementation in real-time, tests and validation are also in our focus.

The control sequence for the conveyor belt example - considering only ON/OFF control - can be developed easily. To describe the control sequence - for this example - flow charts, timing charts and state diagrams can be used.

However, control system design for dynamic systems, where transient response, stability and robustness are important issues, will require additional knowledge and this is the topic on which the book is focused.

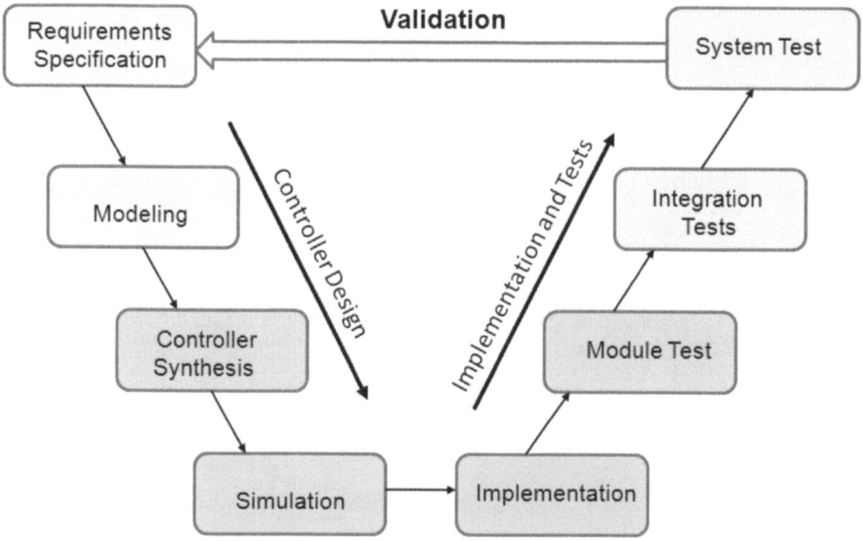

Fig. 2.7 Control algorithm development.

Embedded Software Development

After the control algorithm is designed it is implemented on the embedded hardware. Usually, the control algorithm will be a piece of software code running in real-time. The software development usually follows the V-model presented in Fig. 2.8 and the control engineer plays an active role in testing and integration of the control algorithm.

At this point the control engineer gets in touch with the embedded software world. In software engineering, software reuse is a well-know and widely applied concept.

The software frequently is divided in several abstraction layers as shown in Fig. 2.9. Let assume that the hardware is built around a microcontroller.

The lowest abstraction layer is the microcontroller abstraction layer, also called low level drivers. Usually it contains the following drivers:

- input/output drivers - drivers for analog and digital inputs and outputs (e.g. ADC, DIO, PWM)
- communication drivers - drivers for serial communication (e.g. CAN, LIN)
- memory drivers - drivers for on-chip (internal) and external memory devices (e.g. Flash, EEPROM)
- microcontroller drivers - drivers for internal peripherals (e.g. Watchdog, clock unit).

This layer is a microcontroller vendor specific layer. Some vendors like Infineon provide a graphical user interface (DaVE) to configure and generate the

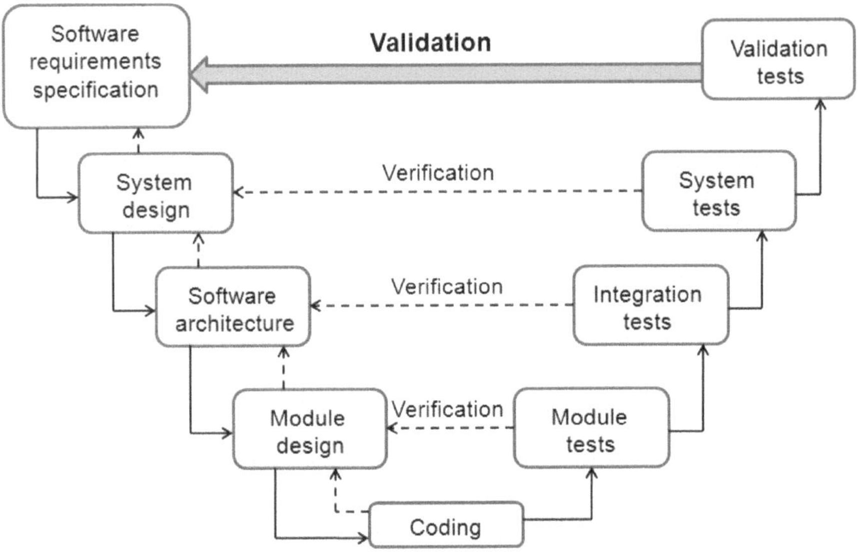

Fig. 2.8 Software development according to V-model.

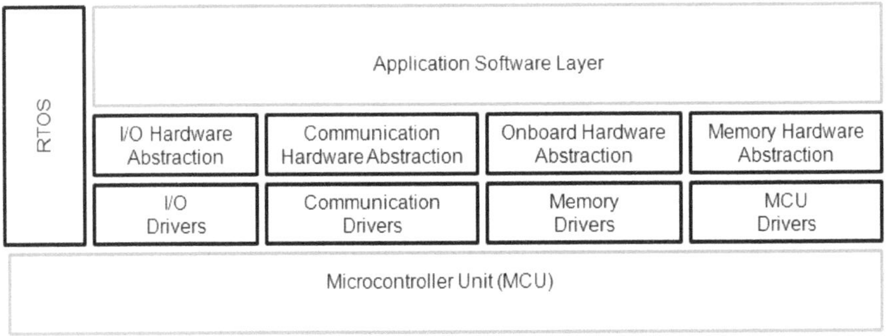

Fig. 2.9 Software layers.

low level drivers, but in most of the cases this software layer is hand coded and frequently reused.

The configuration of ADC channels are done via special function registers (SFRs). We might configure the analog input channel 0 (AD0), with input range from 0V to 5V as:

- 10 bit mode
- sweep mode
- software triggered
- no reference voltage connected.

After the conversion is started, we can retrieve the converted value.

Therefore, we might have C functions like:

- void ADC_Init(void) - all ADC channels are configured here
- void ADC_Start(byte ChannelNum) - this function starts the ADC conversion on the specified channel (ChannelNum)
- uword ADC_GetData0(void) - this function returns the converted analog value on channel 0.

These functions are not detailed here since they are microcontroller specific.

The next layer is the hardware abstraction layer. This layer has the role to hide the microcontroller hardware and layout properties (pin connections and signal level inversions) from higher software layers.

It contains the following parts:

- input/output hardware abstraction - different I/O devices are accessed via an I/O signal interface
- communication hardware abstraction - this layer abstracts from the location of communication controllers
- memory hardware abstraction - abstracts from the location of peripheral memory devices, is on-chip EEPROM and external EEPROM should be accessible via the same mechanism
- onboard device abstraction - contains drivers for onboard devices which cannot be seen as sensors or actuators, like external watchdog.

The hardware abstraction layer abstracts from the microcontroller pin connections but does not abstract from the sensors/actuators.

Let consider an input device, a current sensor type (CS) connected to the analog input channel 0. We assume that the sensor is linear having a measuring range from -50A to 50A. The sensor output is in voltage from 0V to 5V and its characteristics is shown in Fig. 2.10.

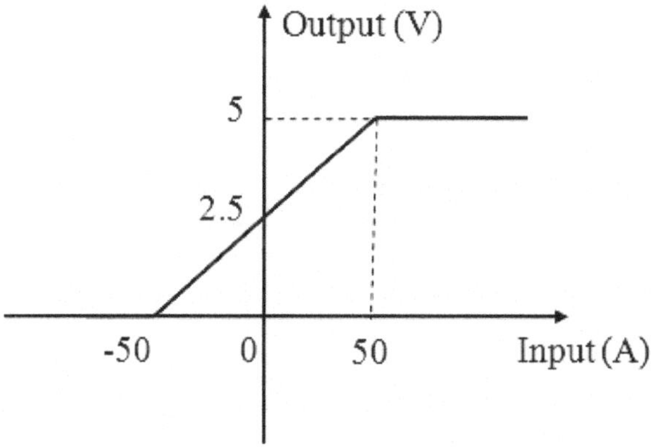

Fig. 2.10 Current sensor characteristics.

For the signal interface first we define the scaling factors mentioned as below:

```
// ADC0 scaling 5V to 1024 (10 bits)
#define ADC0_SCALE    5.0/1024

// current sensor offset 2.5V
#define CS_OFFSET     2.5

// current sensor scaling 50 A to 2,5 V
#define CS_SCALE      50/2.5
```

Then we define a C function Current_CS_Ch0 (), which calls the ADC_GetData0(void) function and returns the measured current value Crt_Ch0. For simplicity - to avoid scaling - floating point arithmetic is used.

```
float Current_CS_Ch0()
{
    float Crt_Ch0;
    uword ADC_Value0;

    // read ADC Ch0
    ADC_Value0 = ADC_GetData0();

    // measured current value in Amps
    Crt_Ch0 = CS_SCALE*(ADC0_SCALE*ADC_Value0 - CS_OFFSET);

return Crt_Ch0;
}
```

The upper software layer is the application software layer, where the control algorithm is implemented. A large part of the embedded code at this software layer can be developed using advanced tools like Matlab/Simulink from MathWorks and can be automatically generated using tools like TargetLink from dSPACE.

For complex applications is strongly recommended to use a real-time operating system (RTOS) for scheduling and prioritizing tasks. Recent market research shows that - in majority of the cases - it is more efficient to buy a commercially available RTOS than to build one.

Finally, about the software reuse, frequently the same hardware platform is used in different control applications - in this case the low level drivers can be reused. We might have the situation that the input devices (sensors) are changed, then the low level drivers and the application software layer can remain unchanged. The control engineer shall be aware about the layered software architecture and software reuse.

Now, let assume that all development phases have been finished and we reach the installation and maintenance phase.

2.1.5 Installation and Maintenance

The first and most important aspect during installation and maintenance is safety. The control system shall be installed according to safety guidelines and rules. The installation manual often contains:

- safety recommendations
- wiring recommendations
- mounting recommendations
- testing and start-up procedures.

Periodic, routine maintenance shall be scheduled for the control system. By checking periodically the critical components and devices, the lifetime of the system can be extended and future problems can be eliminated.

During periodical checks usually it is recommended to:

- check and measure voltages at various points
- measure device current to set a benchmark and compare for changes
- check temperature of different devices
- review diagnostic history, including events and alarms
- perform calibration
- check back-up batteries, and/or replace on a routine schedule
- tighten connections (with the power switched off)
- check alarm systems, emergency stop buttons, indicators, sirens.

In this section, we have briefly reviewed the main requirements for control system design. These are general requirements and are valid for most control systems applied in the industry.

Next, we will narrow our field and we will focus on control system design for dynamical systems, considering a model based approach. Therefore, in the next section we will show:

- how mathematical models are built for control purposes
- how nonlinear models are linearized
- how these models might be used for control system design.

2.2 Mathematical Models for Control

Mathematical models are used in several fields of exact and soft sciences. In engineering design one of the goals is to make a good design based on models. Basically, a model has to be constructed from observed data. There are two main ways of obtaining models, as described next.

2.2.1 Models from Science

A common way of building a model is by making idealizing assumptions and applying the laws of physics, chemistry and so on.

Let us consider a positioning system, which can be described by differential equations based on physical laws. The coefficients in the differential equations will depend on physical constants, such as masses, physical dimensions, etc. These constants can be measured.

The derived model is a so-called white box model, since the structure and parameters of the system (which is modeled) are transparent for the model user.

This method of applying physical laws and taking measurements is successful if the system is not very complex and the governing laws are known.

2.2.2 Models from Experimental Data

The second way of getting a model is by doing experiments on the physical system. Let us consider a real physical system - the plant to be controlled - with one input u and one output y. To design a control system for this plant, we must understand how the input u affects the output y.

The experiment runs like this: apply some input signal $u(t)$, resulting in some output signal $y(t)$. Based on the observed input output pairs (u, y), we can derive a mathematical model. This is a black box model, derived only from input and output data pairs, which does not reveals explicitly the structure and the parameters of the system.

Usually, we know something about the internal structure of the system. Therefore, it might be useful to combine the physical insight into the system (white box model), with the measurement based parameter identification (black box model). We will call this approach grey box model, where the model structure is in accordance with the physical model (applying laws of physics, chemistry, etc) and the parameters are identified by experiments using input and output data pairs (see Fig. 2.11).

Most of the work in control design is in developing a suitable model with proper balance between fidelity and simplicity (see Fig. 2.12), which requires physical insight, advanced system identification techniques and domain expertise.

Regardless of the considered approach no mathematical system can precisely describe a real physical system - there is always uncertainty. Uncertainty means that we cannot predict exactly what the output of a real

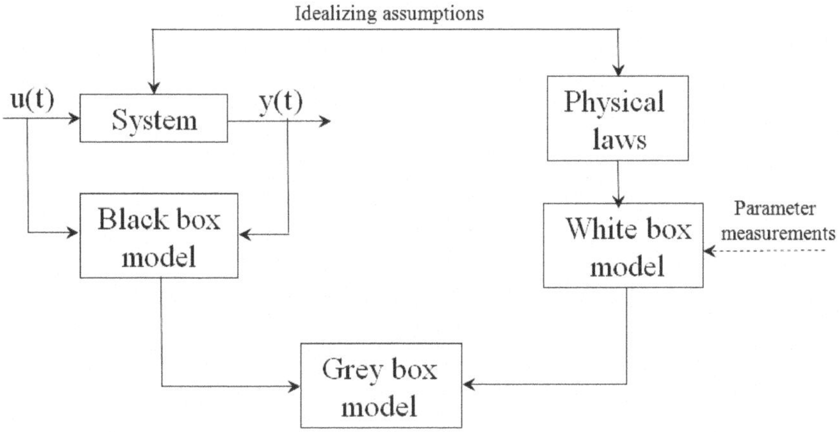

Fig. 2.11 Building grey box models.

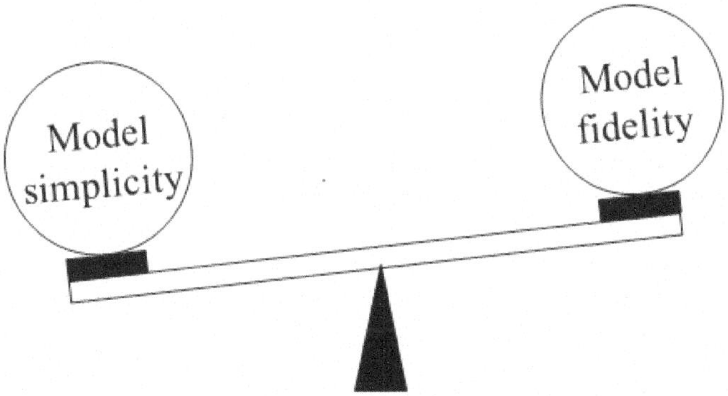

Fig. 2.12 Model simplicity versus model fidelity.

physical system will be even if we know the applied input, so we are uncertain about the system.

Uncertainty arises from two sources:

- unknown or unpredictable inputs such as disturbance, noise, etc.
- unpredictable dynamics.

We will deal with model building for control design purposes, deriving a nominal plant (whenever is possible using a gray box approach) and an upper bound for the plant uncertainty. The relation among the nominal plant, real plant, plant uncertainty and extended plant is shown intuitively in Fig. 2.13.

For a useful control theory (i.e. the controller works properly in a real physical system) the system uncertainty shall be considered during control design.

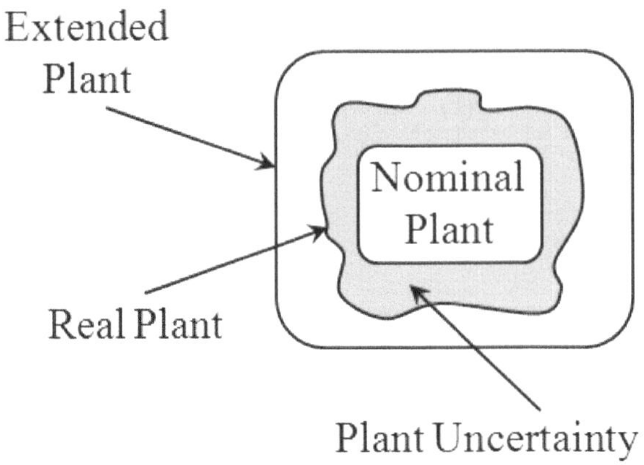

Fig. 2.13 Models for control design.

In this book two approaches for linear controller design are presented:

- classical control approach: the controller is designed for the nominal plant but it is assessed considering the plant uncertainty, too. The controller is designed iteratively until the performance and stability requirements are fulfilled for all possible plants (nominal plant extended with plant uncertainty).
- robust control approach: the controller is designed for the so-called augmented plant (nominal plant extended with plant uncertainty), so the uncertainty is built in explicitly in the model.

One of the main objectives of this book is to present a unified approach for control design, based on system identification and control theory.

The models used here are finite-dimensional, linear and time-invariant. The main reason for this is that they are simplest models for treating fundamental issues in control system design.

The linearity of many mechanical and electrical elements can be assumed over a reasonably large range. However, since in practice we might encounter systems with nonlinearities, we will briefly re-call a few linearization techniques [8, 38, 46].

2.2.3 Linearization of Nonlinear Models

Compensation of Static Nonlinearities

Let us consider a system controlled by a nonlinear valve. The static nonlinearity of the valve is assumed to be:

$$v = f(u) = u^2 \; 0 \leq u \leq 10 \tag{2.1}$$

We would like to compensate this nonlinearity, so we are looking for a function $\hat{f}^{-1}$ such that the product $\hat{f}^{-1}(u)f(u) \approx 10u$ is almost linear. The nonlinear characteristic can be approximated by a linear function $\hat{f}$ using linear interpolation, as shown in Fig. 2.14.

$$\hat{f} = \begin{cases} 6u & 0 \leq u < 6 \\ 16u - 60 & 6 \leq u \leq 10 \end{cases} \tag{2.2}$$

Next, let us assume that the inverse of the linear function $\hat{f}$ is $\hat{f}^{-1}$.

First, let us recall that the inverse of a linear function $y = ax + b$ is $y^{-1} = (x - b)/a$.

Now, we can easily calculate $\hat{f}^{-1}$ the inverse function, however in our example we would like to have the input and the output range defined between 0, ..., 10.

Therefore, the inverse function $\hat{f}^{-1}$ is:

$$\hat{f}^{-1} = \begin{cases} u/6 & 0 \leq u < 3.6 \\ u/16 + 6/16 & 3.6 \leq u \leq 10 \end{cases} \tag{2.3}$$

By simulation we can check that the product $\hat{f}^{-1}(u)f(u)$ is almost linear.

Therefore, the nonlinearity of this system can be compensated by feeding the controller output through the function $\hat{f}^{-1}$ (as shown in Fig. 2.15).

It is left for the reader to simulate this system step response, with and without nonlinear compensation, when the plant transfer function is:

$$P_0(s) = \frac{1}{(s+2)^3} \tag{2.4}$$

and the controller is:

$$K(s) = 0.1 + \frac{1}{2s} \tag{2.5}$$

We can remark that there is improvement in the performance of the closed-loop if the nonlinearity is compensated. By improving the inverse it is possible to make the process even more insensitive to the nonlinearity of the valve. In practice, often is sufficient to approximate the nonlinearity by a few line segments.

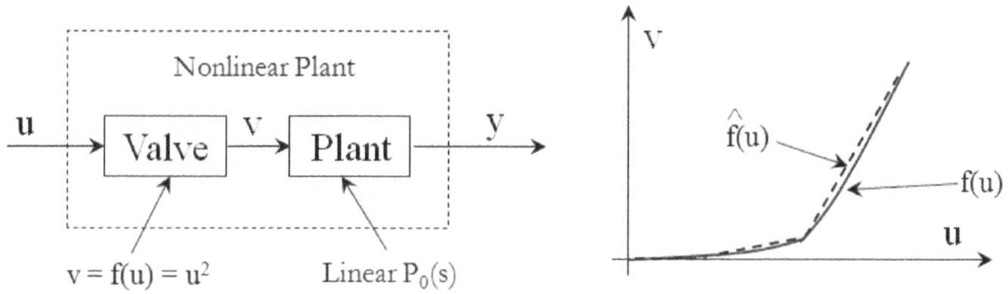

Fig. 2.14 The nonlinear valve characteristic and its two-line approximaion.

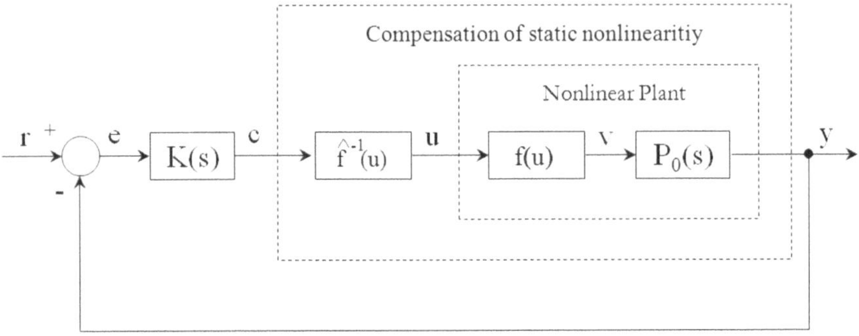

Fig. 2.15 Nonlinear compensation using an approximate inverse.

Linearization around an Operating Point

Let us consider a system described by a nonlinear function

$$y(t) = g(x(t)) \tag{2.6}$$

The normal operating point is designated x_0, then using Taylor series expansion around the operating point we can write:

$$y = g(x) = g(x_0) + \frac{x - x_0}{1!} \frac{\partial g}{\partial x}\bigg|_{x=x_0} + \frac{(x - x_0)^2}{2!} \frac{\partial^2 g}{\partial x^2}\bigg|_{x=x_0} + \ldots \tag{2.7}$$

Neglecting the higher-order terms we can approximate the nonlinear function by a linear one, this is called linear approximation.

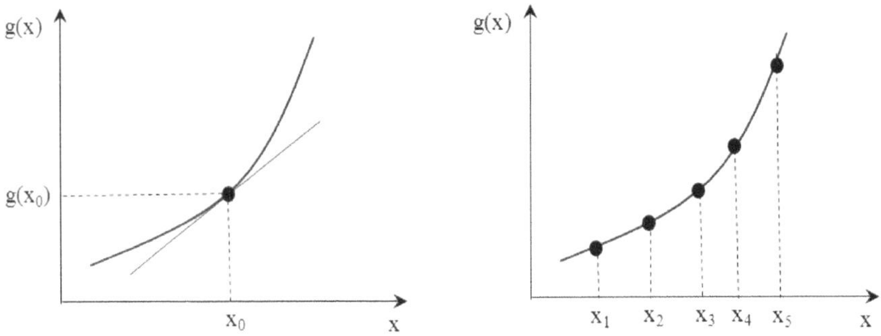

Fig. 2.16 Linearization around operating points.

We can linearize the system around one operating point or around several operating points, as shown in Fig. 2.16. As we will see it later, this is one possible approach to deal with nonlinear systems in a linear framework.

Another approach to linearize nonlinear systems is called feedback linearization and is described in the next section.

Feedback Linearization

Feedback linearization is a method that can be applied to certain systems. The basic procedure is illustrated by an example [46].

Consider the system:

$$\frac{dx_1}{dt} = x_2 + f(x_1) \tag{2.8}$$

$$\frac{dx_2}{dt} = u \tag{2.9}$$

where the function f is nonlinear in x_1, but is differentiable. The first step is to introduce new variables defined as:

$$\begin{aligned} \xi_1 &= x_1 \\ \xi_2 &= x_2 + f(x_1) \end{aligned} \tag{2.10}$$

After derivation equations (2.8) and (2.9) become:

$$\frac{d\xi_1}{dt} = \frac{dx_1}{dt} = x_2 + f(x_1) = \xi_2 \tag{2.11}$$

$$\frac{d\xi_2}{dt} = \frac{dx_2}{dt} + f'(x_1) = u + f'(\xi_1) \tag{2.12}$$

By introducing the control law:

$$u = -a_2\xi_1 - a_1\xi_2 - f'(\xi_1) + v \tag{2.13}$$

we obtain a linear closed-loop system described by:

$$\frac{d}{dt}\begin{bmatrix} \xi_1 \\ \xi_2 \end{bmatrix} = \begin{bmatrix} 0 & 1 \\ -a_2 & -a_1 \end{bmatrix}\begin{bmatrix} \xi_1 \\ \xi_2 \end{bmatrix} + \begin{bmatrix} 0 \\ 1 \end{bmatrix} v \tag{2.14}$$

By transforming back to original coordinates the control law can be written as:

$$u = -a_2x_1 - (a_1 + f'(x_1))(x_2 + f(x_1)) + v \tag{2.15}$$

The closed-loop system obtained in this example will behave like a linear system. This is the reason why the method is called feedback linearization.

The benefit of feedback linearization is that a linear compensator can be designed for the linearized plant (using well-established linear control design techniques), as shown in Fig. 2.17. However, there are disadvantages such as: the feedback linearization requires exact knowledge of the nonlinearity since exactly cancels the nonlinear terms (is not a very robust scheme). Furthermore, real-time implementation of the feedback linearization law is computationally intensive and might be numerically fragile.

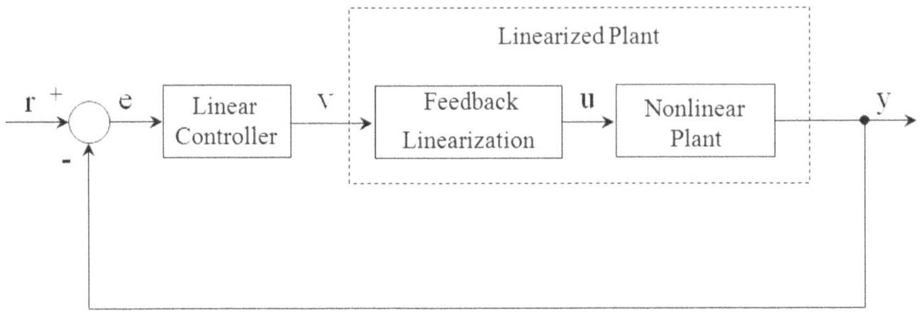

Fig. 2.17 Feedback linearization.

Another similar approach, which is used for linearization of nonlinear systems is called backstepping, for details the interested reader should consult [8, 46].

As mentioned earlier, models we are going to consider are finite-dimensional, linear and time-invariant. The resulting design techniques are suitable for a large class of engineering problems, since most systems are built to be as close as possible to linear time-invariant systems so that they are more easily controlled [25].

In the next section we are going to describe the main characteristics of feedback control systems. We will emphasize the advantages of a feedback control system but we will also re-call the limitations of control systems. At the end we briefly summarize the concepts of stability and relative stability.

2.3 Control System's Characteristics

One of the most important characteristics of control systems is their transient response. If an open-loop control system does not provide a satisfactory response, then the process, $P(s)$ must be replaced with a suitable process. By contrast, a closed-loop system can often be adjusted to yield the desired response by adjusting the feedback loop parameters.

Another important effect of feedback in control systems is the control and partial elimination of the effect of disturbance signals. Many control systems are subject to disturbance signals. Feedback systems have the beneficial aspect that the effect of distortion, noise and unwanted disturbances can be effectively reduced.

The utilization of feedback in control systems is primarily to reduce the sensitivity of the system to parameter variations and the effect of disturbance inputs. A feedback system is valuable because it provides the engineer with the ability to adjust the transient response.

Let us consider the control structure with unity feedback shown in Fig. 2.18 where:

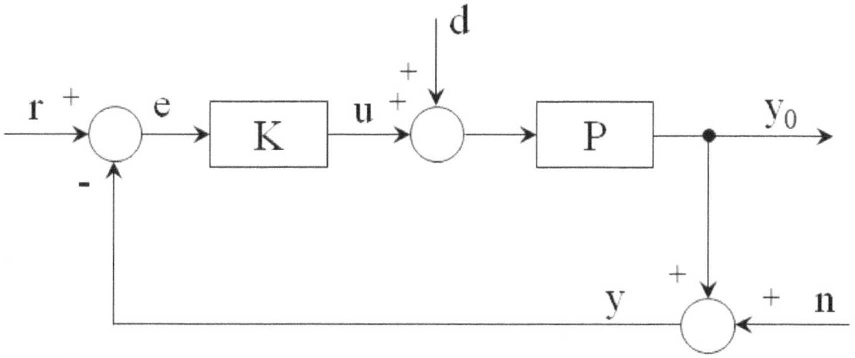

Fig. 2.18 Block diagram of a feedback system.

$K(s)$ is the controller transfer function
$P(s)$ is the plant transfer function
r is the control reference
d is the external disturbance
n is the measurement noise
y_0 is the controlled magnitude
e is the control error, which by definition is $e = r - y_0$
u is the control signal

The open-loop transfer function is noted by: $L(s) = K(s)P(s)$. The transfer function from the control reference r to the control error e is noted by $S(s)$:

$$S(s) = \frac{1}{1 + K(s)P(s)} = \frac{1}{1 + L(s)} \tag{2.16}$$

and is called sensitivity transfer function.

The transfer function from the control reference r to the control output y_0 is noted by $T(s)$:

$$T(s) = \frac{K(s)P(s)}{1 + K(s)P(s)} = \frac{L(s)}{1 + L(s)} \tag{2.17}$$

and is called complementary sensitivity transfer function.

We remark that:

$$S(s) + T(s) = 1 \tag{2.18}$$

Next, let us briefly present the main characteristics of the control systems.

2.3.1 Disturbance Attenuation

In this case $d(t) \neq 0$, $r(t) = 0$ and let us assume $n(t) = 0$. In this situation, the magnitude of the output, $y(t)$, should be as small as possible, closed to the desired response $r(t)$. Using transfer function representations we can write:

$$Y(s) = \frac{P(s)}{1 + K(s)P(s)} D(s) + \frac{K(s)P(s)}{1 + K(s)P(s)} R(s) \tag{2.19}$$

So the response due to $d(t)$ is:

$$Y(s) = \frac{P(s)}{1 + K(s)P(s)} D(s) \tag{2.20}$$

For good disturbance attenuation we need to make $1/(1 + K(s)P(s))$ small by an appropriate choice of $K(s)$.

Let $|D(j\omega)|$ be the magnitude of the disturbance in frequency domain Ω, and let us suppose:

$$\begin{cases} |D(j\omega)| \neq 0 \ for \ \omega \in \Omega \\ |D(j\omega)| \approx 0 \ for \ \omega \notin \Omega \end{cases} \tag{2.21}$$

If the controller is designed in such a way that

$$|1 + K(j\omega)P(j\omega)|^{-1} << 1 \ for \ \omega \in \Omega \tag{2.22}$$

then good disturbance attenuation is achieved by feedback.

2.3.2 Tracking

Now consider the dual problem $r(t) \neq 0$, $d(t) = 0$ and $n(t) = 0$. In this case the tracking error $e(t) = r(t) - y(t)$ should be as small as possible.

The tracking error is:

$$E(s) = \frac{1}{1 + K(s)P(s)} R(s) \tag{2.23}$$

Therefore, similar to disturbance attenuation, one should select $K(s)$ in such a way that:

$$|1 + K(j\omega)P(j\omega)|^{-1} << 1 \tag{2.24}$$

in the frequency region where $|R(j\omega)|$ is large.

Next, let us make a few remarks about the bandwidth of the closed-loop system. Usually, the bandwidth of the closed-loop system is noted by ω_B and is a measure of a system's ability to faithfully reproduce an input signal.

The bandwidth is the frequency ω_B at which the frequency response has declined 3 dB (see Fig. 2.19). The speed of response to step input will be roughly proportional to ω_B and the settling time is inversely proportional to ω_B.

Moreover, it may be empirically shown that the crossover frequency on the open-loop system noted by ω_C (where $|L(j\omega_C)| = 0$) is related to the closed-loop system bandwidth, ω_B by the approximation:

$$\omega_B \approx 1.6 \cdot \omega_C \; for \; 0.2 \le \xi \le 0.8 \tag{2.25}$$

where ξ is the damping ratio of the closed-loop system.

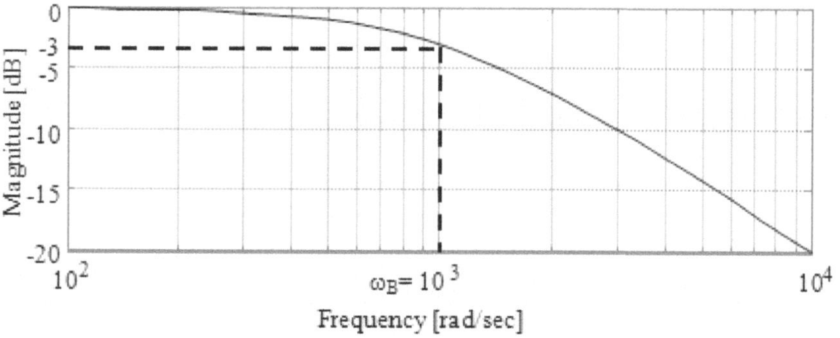

Fig. 2.19 Control system's bandwidth.

2.3.3 Sensitivity to Parameter Variations

Let us assume that due to parameter variations of the plant P is changing to $P' = P + \Delta P$ then the open-loop transfer function $L = KP$ is changing as $L' = L + \Delta L$ and the closed-loop transfer function becomes $T' = T + \Delta T$.

The sensitivity of a function F, which depends on a parameter α is defined as:

$$S_\alpha^F = \lim_{\Delta\alpha \to 0} \frac{\Delta F/F}{\Delta\alpha/\alpha} = \frac{\alpha}{F}\frac{\partial F}{\partial \alpha} \tag{2.26}$$

Applying the definition, the sensitivities of $L(s)$ and $T(s)$ due to variations in P are:

$$S_P^L = \lim_{\Delta P \to 0} \frac{\Delta L/L}{\Delta P/P} = 1 \tag{2.27}$$

$$S_P^T = \lim_{\Delta P \to 0} \frac{\Delta T/T}{\Delta P/P} \tag{2.28}$$

Then we can write:

$$\Delta T = T' - T = \frac{L'}{1+L'} - \frac{L}{1+L} = \frac{\Delta L}{(1+L')(1+L)} \tag{2.29}$$

as well as:

$$\frac{\Delta T}{T} = \frac{\Delta L}{L} \frac{1}{1+L'} \tag{2.30}$$

then we can write:

$$S_P^T = S_P^L \frac{1}{1+L'} \approx S_P^L \frac{1}{1+KP} \tag{2.31}$$

The previous equation shows that the sensitivity of the closed-loop is attenuated by $1+KP$ times compared with the sensitivity of the open-loop. Therefore, the effect of parameter variations in P can be made small using feedback control, the same cannot be achieved by open-loop control.

2.3.4 Control System's Limitation

From the tracking control problem, neglecting the external noise, we have seen that for frequencies where $|L(j\omega)| >> 1 \Rightarrow T(s) \approx 1$ we can obtain good tracking performance. However, the control signal in that frequency range can be written as:

$$U(s) = \frac{K}{1+KP} R(s) = \frac{T(s)}{P(s)} R(s) \approx \frac{1}{P(s)} R(s) \tag{2.32}$$

The previous equation shows that for higher frequencies (where usually $|P(j\omega)|$ is low) the amplitude of the control signal will be relatively large, leading to significant component wear. Furthermore, in practice the control signal is always limited, which impose limitations on the control system. Therefore, there are limitations how big $|L(j\omega)|$ can be made.

Next, let us consider the influence of measurement noise and let us neglect the external disturbance $d(t) = 0$. The tracking error is defined by: $e(t) = r(t) - y_0(t)$, then it is possible to prove (this is left to the reader) that the control error is:

$$E(s) = \frac{1}{1+KP} R(s) + \frac{KP}{1+KP} N(s) = S(s)R(s) + T(s)N(s) \tag{2.33}$$

For good tracking and noise reduction we desire:

$$\begin{aligned} |S(j\omega)R(j\omega)| << 1 \ \forall \ \omega \\ |T(j\omega)N(j\omega)| << 1 \ \forall \ \omega \end{aligned} \tag{2.34}$$

This is equivalent with: $|S(j\omega)|$ should be small in the frequency region where $|R(j\omega)|$ is large and $|T(j\omega)|$ should be small where $|N(j\omega)|$ is large.

Since,

$$S(s) + T(s) = 1 \tag{2.35}$$

it is impossible to make both $|S(j\omega)|$ and $|T(j\omega)|$ small at the same frequency ω. Fortunately, in most practical cases $|N(j\omega)|$ is small when $|R(j\omega)|$ is large and vice versa.

There are also other constraints on the sensitivities, shown by Bode and known as Bode's integrals [73].

$$\int_0^\infty log|S(j\omega)|d\omega = \pi \sum p_i \tag{2.36}$$

$$\int_0^\infty log|T(1/j\omega)|d\omega = \pi \sum \frac{1}{z_i} \tag{2.37}$$

where p_i are the right half plane poles of $L(s) = P(s)K(s)$ and z_i are the right half plane zeros of $L(s)$.

Bode's integrals are conservation laws. They state precisely that the integrated value of the *log* of the magnitude of the sensitivity function is conserved under the action of feedback.

The total amount of this quantity is always the same. It is equal to zero for stable plant/compensator pairs, and it is equal to some fixed positive amount for unstable ones. Sensitivity improvements in one frequency range must be paid for the sensitivity deteriorations in another frequency range, and this price is higher if the plant is open-loop unstable.

These equations imply that the sensitivities can be made small at one frequency only at the expense of increasing the sensitivity at other frequencies (see Fig. 2.20). This phenomenon is sometimes called the water-bed effect. It also follows from the equations that the presence of poles in the right half plane increase the sensitivity and the zeros in the right half plane increase the complementary sensitivity.

An argument is sometimes made that the Bode's integrals are not really restrictive because we only seek to make the sensitivity function small over a finite frequency band, then we have an infinite frequency range left over, so the Bode's integrals are easily satisfied.

The weakness of this argument is that physical systems do not exhibit good frequency fidelity beyond a certain bandwidth. This is due to uncertain or unmodelled dynamics in the plant, to digital control implementations, to power limitations, to nonlinearities, and to many other factors.

Let us call that bandwidth the 'available bandwidth', and let us note with ω_a, to distinguish it from other bandwidths such as crossover ω_C or 3 dB magnitude loss ω_B. The available bandwidth is the frequency up to which we can keep $K(j\omega)P(j\omega)$ close to a nominal design and beyond which we can only guarantee that the actual loop magnitude will attenuate rapidly enough.

In today's popular robust control jargon, the available bandwidth is the frequency range over which the unstructured multiplicative perturbations are substantially less than unity. Let us note that the available bandwidth is not

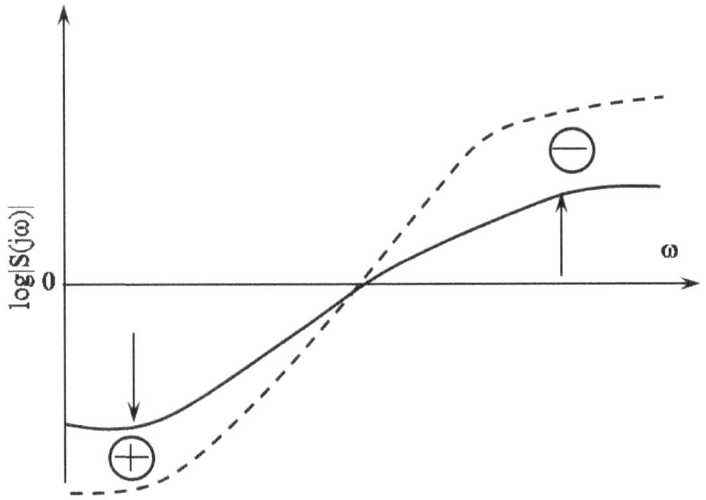

Fig. 2.20 Bode's integral - infinite bandwidth.

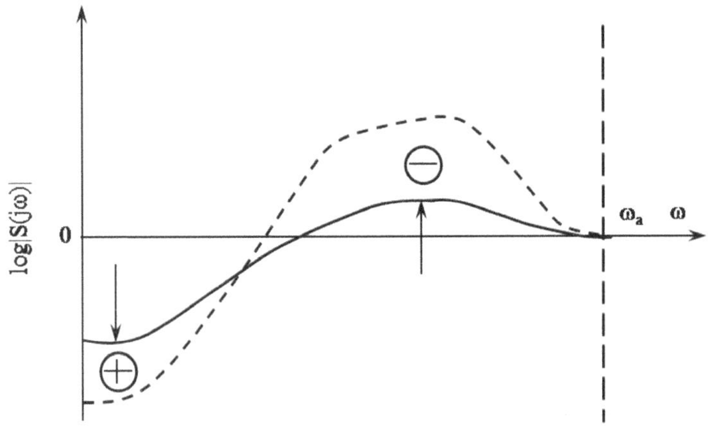

Fig. 2.21 Bode's integral - finite bandwidth.

a function of the compensator or of the control design process, rather it is an a priori constraint imposed by the physical hardware we use in the control loop. Most importantly the available bandwidth is always finite.

Given all this, Bode's integrals are reduced to finite integrals over the range $0 \leq \omega \leq \omega_a$:

$$\int_0^{\omega_a} log|S(j\omega)|d\omega = \pi \sum p_i \qquad (2.38)$$

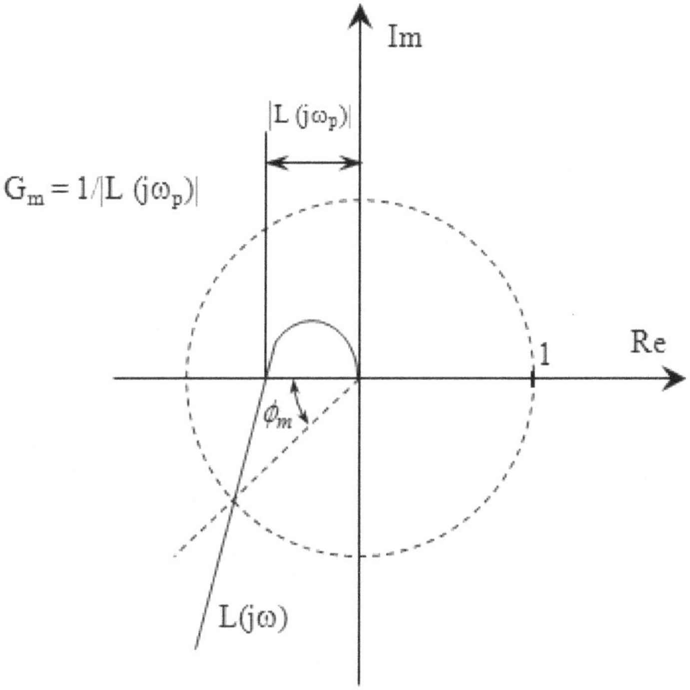

Fig. 2.24 Nyquist curve with gain and phase margins.

A gain margin G_m implies that the gain can be increased with a factor less than G_m, without making the system unstable. Similarly, for a system with a phase margin ϕ_m it is possible to increase the phase shift in the loop by a quantity less than ϕ_m without making the system unstable.

After reviewing the main characteristics of feedback control systems, the last section of this chapter is dedicated to performance specifications.

2.4 Performance Specifications for Linear Systems

The performance specifications of linear systems usually are defined in terms of settling time and percent of overshoot, however in many cases the control design is performed in frequency-domain. Therefore, it is important to know how to translate performance specifications given in time-domain into performance specifications is frequency-domain.

Furthermore, in practice the designed control system is often required to behave like a second-order system, having a well-defined settling time T_{set} and percent of overshoot $P.O.$, shown in Fig. 2.26.

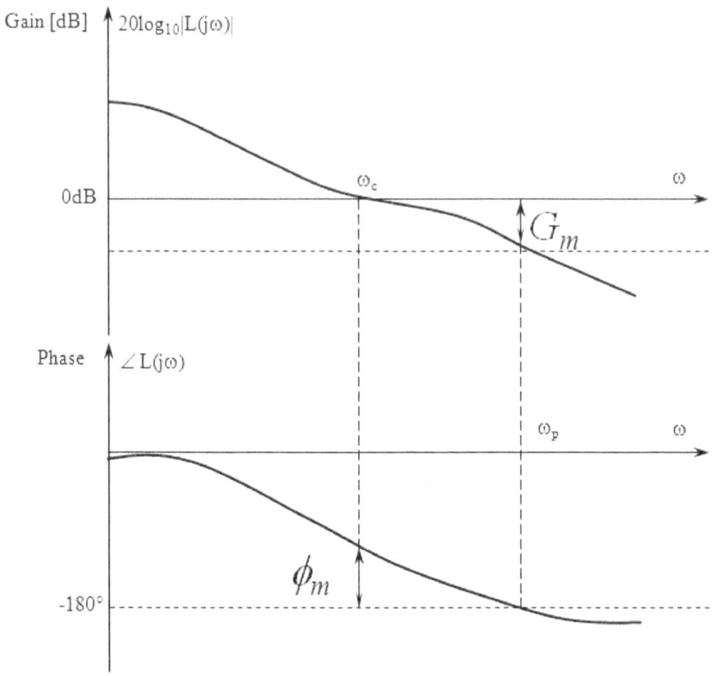

Fig. 2.25 Gain and phase margins - Bode plots.

Therefore, let us consider that the control system behaves like a second-order system, then we can rewrite:

$$Y(s) = \frac{\omega_n^2}{s^2 + 2\xi\omega_n s + \omega_n^2} R(s) \tag{2.52}$$

where ξ is the damping and ω_n is the natural frequency.

With a unit step input $R(s) = 1/s$ we obtain:

$$Y(s) = \frac{\omega_n^2}{s(s^2 + 2\xi\omega_n s + \omega_n^2)} \tag{2.53}$$

In the case of second-order systems the settling time T_{set} is approximately:

$$T_{set} \approx \frac{4}{\xi\omega_n} \tag{2.54}$$

and the percent of overshoot is:

$$P.O. = 100e^{-\xi\pi/\sqrt{1-\xi^2}} \tag{2.55}$$

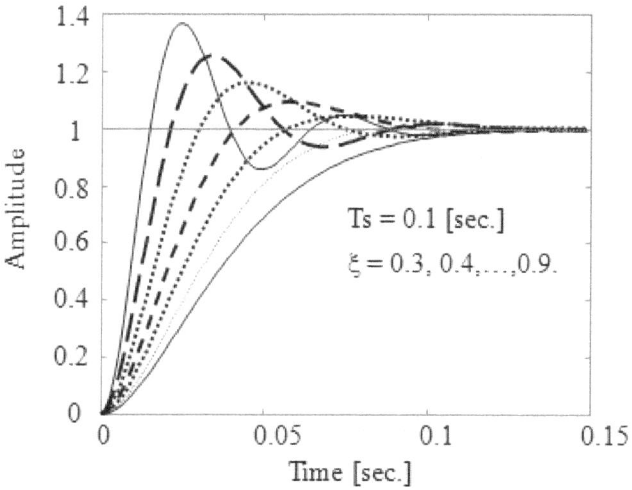

Fig. 2.26 Step response - second-order system.

Table 2.1 Performance of a second-order system.

Damping ratio [ξ]	0.9	0.8	0.7	0.6	0.5	0.4	0.3
Percent of overshoot [P.O.]	0.2	1.5	4.6	9.5	16.3	25.4	37.2

Based on the results summarized in Table 2.1 it is possible to compute the damping ratio ξ as well as the natural frequency ω_n of the closed-loop transfer function.

Let us require $T_{set} = 0.5$ and P.O. $= 4\%$ then we can easily compute the damping ratio and the natural frequency:

$$\xi = 0.707 \; \omega_n = 8\sqrt{2} \tag{2.56}$$

The poles of the closed-loop $T(s)$ are:

$$p_{1,2} = -\xi\omega_n \pm j\omega_n\sqrt{1-\xi^2} \tag{2.57}$$

Therefore, the poles of $T(s)$ should lie in a region of the complex plane defined by the minimum allowable ξ and $\xi\omega_n$, the dashed area shown in Fig. 2.27.

This background information will be useful for us to define the performance specifications in frequency domain.

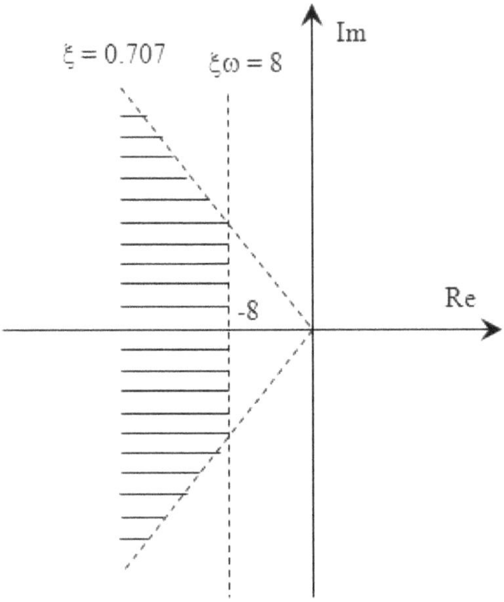

Fig. 2.27 Region of the desired closed-loop poles.

Chapter 3

System Identification and Model-Order Reduction

'Whether one succeeds or fails depends on one's thought and attitude.' — Chung, Ju Yung, Founder of Hyundai Heavy Industries

Summary:

This chapter is dedicated to system identification and model-order reduction problems. The following items are discussed:

- the main steps of the system identification procedure
- concepts like "persistence of excitation" and "informative system identification experiment"
- input signal design for system identification experiments
- model-order reduction methods based on balanced truncation techniques
- nominal plant and plant uncertainties
- the system identification procedure is illustrated in the case of a brushless d.c. drive as well as a fuel cell.

All controller design techniques presented here are model based. Therefore, a mathematical model of the plant must be derived using for example system identification techniques.

From control viewpoint, the goal of system identification is to derive a nominal plant model and a bound for the plant uncertainty. In some cases, the plant is treated as a black box, assuming it is linear and an input signal is applied to obtain the system response. The collected data may be noisy, so the 'best fit' may lead to a high-order model.

The common practice is to find a low-order model that explains the data 'reasonably well'. The difference between the actual output and the low-order model response can be seen as the response of the uncertain part of the plant. Furthermore, the derived model uncertainty must be taken into account during the controller design, for example, in the framework of the robust control theory.

In addition to the plant uncertainty identified using system identification techniques, we must take into account parameter variations due to manufacturing dispersion.

The techniques presented in this chapter assume that the system is linear or can be approximated by a linear system around the operating point. Furthermore, the parameters of the system - in most of the cases - varies with the temperature, which must be investigated. Therefore, it might be the case that the experiments shall be repeated around different operating points as well as at different temperatures.

3.1 Model Building and Model Structures

System identification deals with the problem of building mathematical models of dynamical systems based on observed data from the system. During the system identification process we will note the system's input and output at time t by $u(t)$ and $y(t)$ respectively [51].

Perhaps the most basic relationship between the input and output is the linear difference equation:

$$y(t) + a_1 y(t-1) + \cdots + a_n y(t-n) = b_1 u(t-1) + \cdots + b_m u(t-m) \quad (3.1)$$

We have chosen to represent the system in discrete time, since observed data are always collected by sampling. We assume the sampling interval to be one time unit to make notation simpler.

A pragmatic way to see equation (3.1) is to view it as a way of determining the next output value given previous observations:

$$y(t) = -a_1 y(t-1) - \cdots - a_n y(t-n) + b_1 u(t-1) + \cdots + b_m u(t-m) \quad (3.2)$$

For a more compact notation we introduce the vectors:

$$\begin{aligned} \theta &= [a_1 \quad \dots \quad a_n \quad b_1 \quad \dots \quad b_m]^T \\ \varphi(t) &= [-y(t-1) \dots -y(t-n) \; u(t-1) \dots u(t-m)]^T \end{aligned} \quad (3.3)$$

With these, equation (3.2) can be written as:

$$y(t) = \varphi^T(t)\theta \quad (3.4)$$

To emphasize that the calculation of the $y(t)$ is from past data we will write:

$$\widehat{y}(t) = \varphi^T(t)\theta \quad (3.5)$$

Now suppose for a given system that we don't know the values of the parameters in θ, but we have recorded inputs and outputs over the time interval: $1 \leq t \leq N$.

$$Z^N = \{u(1), y(1), ..., u(N), y(N)\} \quad (3.6)$$

An obvious approach is to compute θ based on the least squares method, in this order we have to minimize the following function:

$$V_N(\theta, Z^N) = \frac{1}{N}\sum_{t=1}^{N}(y(t) - \widehat{y}(t))^2 = \frac{1}{N}\sum_{t=1}^{N}(y(t) - \varphi^T(t)\theta)^2 \tag{3.7}$$

Since $V_N(\theta, Z^N)$ is quadratic we can find the minimum value easily, by setting the derivative to zero,

$$\frac{d}{d\theta}V_N(\theta, Z^N) = \frac{2}{N}\sum_{t=1}^{N}\varphi(t)\left(y(t) - \varphi^T(t)\theta\right) = 0 \tag{3.8}$$

which gives:

$$\sum_{t=1}^{N}\varphi(t)y(t) = \sum_{t=1}^{N}\varphi(t)\varphi^T(t)\theta \tag{3.9}$$

or

$$\hat{\theta} = \left[\sum_{t=1}^{N}\varphi(t)\varphi^T(t)\right]^{-1}\sum_{t=1}^{N}\varphi(t)y(t) \tag{3.10}$$

Once the vector $\varphi(t)$ is defined, the solution can easily be computed by modern software tools. It is important to remark that the estimate has a unique solution iff the matrix:

$$\sum_{t=1}^{N}\varphi(t)\varphi^T(t) \tag{3.11}$$

has full rank - condition analyzed in the next sections. The model structure presented in equation (3.1) is linear in θ and is known in statistics as linear regression. Furthermore, the model structure has the standard name ARX-model - Auto-Regressive-eXogeneous model.

As conclusion, the construction of a model from data involves three basic entities:

- a data set, like Z^N
- a set of candidate models, basically a model structure
- a rule by which candidate models can be assessed using the data like least squares method (LSM).

To collect the data, a system identification experiment shall be designed, which includes several choices, such as which signals to measure and when to measure them and which signals to manipulate and how to manipulate them. We will focus on this topic in the next section but first let us review possible model structures.

3.1.1 Model Structures

The system identification loop has a natural logical flow (shown in Fig. 3.1); first collect data, then choose a model set; then pick the "best" model in this set. It is quite likely, that the model first obtained will not pass the model validation test.

Furthermore, the model can be regarded as a good description of certain aspects of the system but can never be accepted as a final and true description of the system [51].

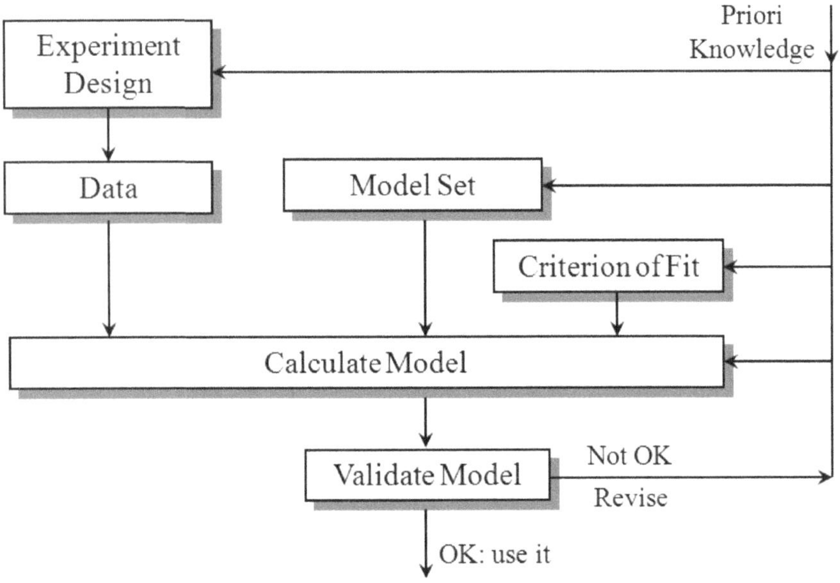

Fig. 3.1 The system identification loop.

The most common description for a linear system with additive disturbance presented in Fig. 3.2 is:

$$y(t) = P(z^{-1})u(t) + H(z^{-1})w(t) \tag{3.12}$$

where $w(t)$ is assumed to be a sequence of independent random variables with zero mean and variance λ and $H(z^{-1})$ is the noise model, also is assumed that $P(z^{-1})$ is stable.

It can be proved that the stochastic process described by $v(t) = H(z^{-1})w(t)$, where $w(t)$ is a sequence of independent random variables with zero mean value and covariance λ, has the spectrum [51]:

$$\Phi_v(\omega) = \lambda \mid H(e^{j\omega}) \mid^2 \tag{3.13}$$

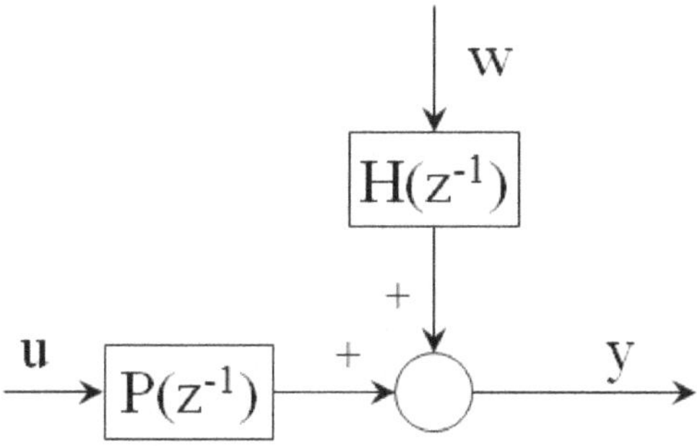

Fig. 3.2 Linear system with additive disturbance.

Furthermore, if a stationary process $v(t)$ has rational spectrum $\Phi_v(\omega)$, we can represent it as:

$$v(t) = T(z^{-1})w(t) \tag{3.14}$$

and let us note:

$$T(z^{-1}) = \frac{C(z^{-1})}{A(z^{-1})} \tag{3.15}$$

The model structures may rise to 32 different model sets, depending on which of the five polynomials $A(z^{-1}), B(z^{-1}), C(z^{-1}), D(z^{-1})$ and $F(z^{-1})$ are used:

$$A(z^{-1})y(t) = \frac{B(z^{-1})}{F(z^{-1})}u(t) + \frac{C(z^{-1})}{D(z^{-1})}w(t) \tag{3.16}$$

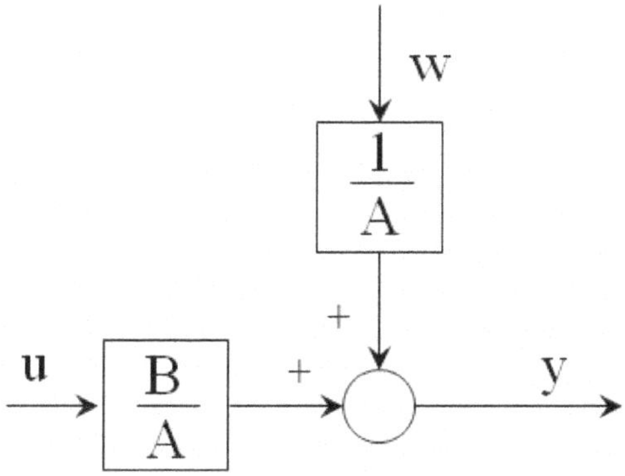

Fig. 3.3 ARX model structure.

The previous structure (3.16) is too general for most practical purposes. One or several of the five polynomials would be fixed to unity in applications.

Some common black-box single input-single output models are special cases of the previous equations and are summarized in the Table 3.1 and some presented in Fig. 3.3, Fig. 3.4 and Fig. 3.5.

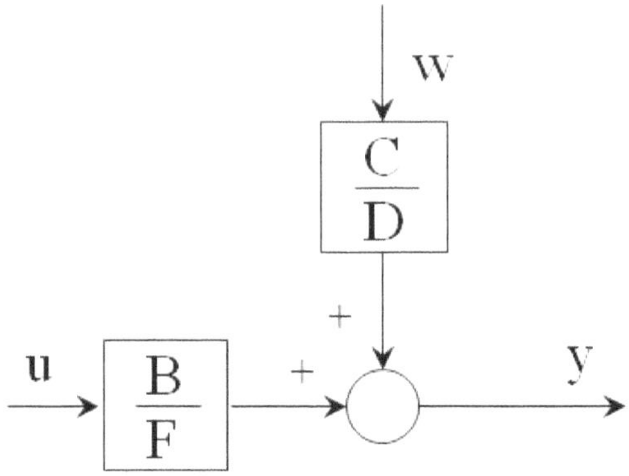

Fig. 3.4 The BJ model structure.

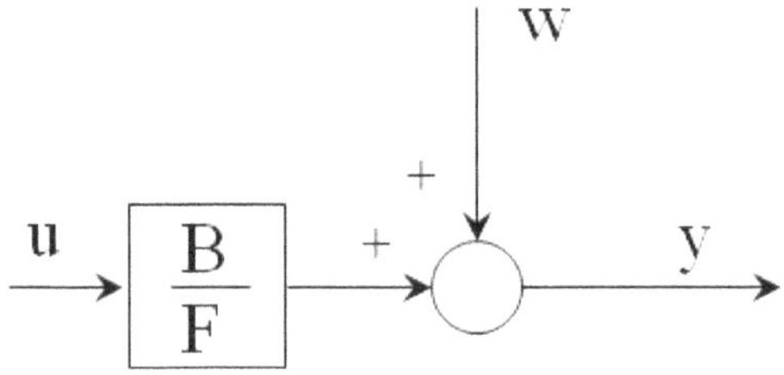

Fig. 3.5 Output error model structure.

During our investigations we will focus on the ARX models, their structure has been already introduced (see equation (3.1)). Furthermore, we can denote the model structure by $\mathcal{M}$, while a particular model corresponding to the parameter value θ is denoted by $\mathcal{M}(\theta)$ such a parametrization is instrumental in conducting a search for "best models". Two different philosophies may guide the choice of a parametrized model sets:

- black-box model structures - the prime idea is to obtain flexible model sets that can accommodate a variety of systems without looking into their internal structures, the input-output model structures are of this character

3.2.1 Requirements Imposed on the Input Signal

At this point we have to introduce new concepts like:

- persistently exciting signal of order n
- informative system identification experiments.

Definition 3.1. *A quasi-stationary signal $u(t)$ with spectrum $\Phi_u(\omega)$ is said to be* **persistently exciting** *of order n if for all filters of the form:*

$$M_n(z^{-1}) = m_1 z^{-1} + \cdots + m_n z^{-n} \tag{3.27}$$

the relation

$$\mid M_n(e^{j\omega}) \mid^2 \Phi_u(\omega) \equiv 0 \tag{3.28}$$

implies that $M_n(e^{j\omega}) \equiv 0$.

We also notice that $|M_n(e^{j\omega})|^2 \Phi_u(\omega)$ is the spectrum of the signal $v(t) = M_n(z^{-1})u(t)$. Hence a signal that is persistently exciting of order n cannot be filtered to zero by an $(n-1)$th-order moving-average filter.

Lemma 3.1. *Let $u(t)$ be a quasi-stationary signal and let the $n \times n$ matrix R_n be defined by*

$$R_n = \begin{bmatrix} \phi_{uu}(0) & \phi_{uu}(1) & \ldots & \phi_{uu}(n-1) \\ \phi_{uu}(1) & \phi_{uu}(0) & \ldots & \phi_{uu}(n-2) \\ . & . & \ldots & . \\ . & . & \ldots & . \\ . & . & \ldots & . \\ \phi_{uu}(n-1) & \phi_{uu}(n-2) & \ldots & \phi_{uu}(0) \end{bmatrix} \tag{3.29}$$

where ϕ_{uu} is the auto-correlation function, then $u(t)$ is persistently exciting of order n iff R_n is nonsingular, that means $\det R_n \neq 0$

Example 3.3. *Let us consider a sinusoidal signal $u(t) = \sin \omega t$. It is easy to prove that:*

$$\phi_{uu}(\xi) = \frac{1}{2} cos \omega \xi \tag{3.30}$$

Then, we compute:

$$\det R_1 = \phi_{uu}(0) = \frac{1}{2} \neq 0 \tag{3.31}$$

$$\det R_2 = \frac{1}{2} det \begin{bmatrix} 1 & cos\omega \\ cos\omega & 1 \end{bmatrix} \neq 0 \tag{3.32}$$

$$\det R_3 = \frac{1}{2} det \begin{bmatrix} 1 & cos\omega & cos2\omega \\ cos\omega & 1 & cos\omega \\ cos2\omega & cos\omega & 1 \end{bmatrix} = 0 \tag{3.33}$$

Therefore, the signal $u(t) = \sin \omega t$ is persistently exciting of order two, $\det R_2 \neq 0$.

Next, consider for example, the $u(t)$ signal consisting of n different sinusoids:

$$u(t) = \sum_{k=1}^{n} u_k cos(\omega_k t) \tag{3.34}$$

each sinusoid gives rise to a spectral line at ω_k and $-\omega_k$. This signal is persistently exciting of order $2n$. If one of the frequencies equals 0, the order drops to $2n - 1$, since this only gives one spectral line. Similarly, if one of the frequencies equals π the order drops by (another) 1.

Theorem 3.1. *Consider a set of single-input, single-output models $\mathcal{M}$ given by the rational transfer functions:*

$$G(z^{-1}) = \frac{b_1 z^{-1} + \cdots + b_m z^{-m}}{1 + a_1 z^{-1} + \cdots + a_n z^{-n}} = \frac{B(z^{-1})}{A(z^{-1})} \tag{3.35}$$

Then an open-loop experiment with an input that is persistently exciting of order $n + m$ is **informative** *with respect to $\mathcal{M}$.*

Proof

The proof of this theorem is performed by 'reductio ad absurdum'.

Suppose that the input signal $u(t)$ persistently exciting of order $n + m$ is non-informative with respect to $\mathcal{M}$. Then exist two different models denoted by $G_1(z^{-1})$ and $G_2(z^{-1})$, such that:

$$G_1(z^{-1}) \neq G_2(z^{-1}) \tag{3.36}$$

Let us note with ΔG the difference between the two models:

$$\Delta G(z^{-1}) = G_1(z^{-1}) - G_2(z^{-1}) = \frac{B_1(z^{-1})A_2(z^{-1}) - B_2(z^{-1})A_1(z^{-1})}{A_1(z^{-1})A_2(z^{-1})} \tag{3.37}$$

where the numerator of the $\Delta G(z^{-1})$ is a polynomial of degree at most $n+m$.

From the definition of the persistence of excitation follows that:

$$|\Delta G(e^{j\omega})|^2 \Phi_u(\omega) \equiv 0 \textit{ implies that } \Delta G(e^{j\omega}) \equiv 0 \tag{3.38}$$

which means that $G_1(z^{-1}) = G_2(z^{-1})$, which is in the contradiction with the initial assumption.

Therefore, the input $u(t)$ persistently exciting of order $n+m$ is informative with respect to $\mathcal{M}$.

Remark 3.1. *An open-loop experiment is* **informative** *if the input is persistently exciting. The necessary order of persistent excitation equals the number of parameters to be estimated in this case.*

3.2.2 Input Signal Design

The requirement from the previous section that the data should be informative means for open-loop operation that the input should be persistently exciting (p.e.) of a certain order, that it contains sufficiently distinct frequencies. This section discusses good and typical choices of input signals [51].

For the identification of linear systems, there are three basic facts that govern the choices:

- the asymptotic properties of the estimate (bias and variance) depend only on the input spectrum not the actual waveform of the input;
- the input must have limited amplitude $|u(t)| \leq \bar{u}$;
- periodic inputs may have certain advantages.

The desired property of the waveform is defined in terms of the **crest factor**, which for a zero mean signal is defined as:

$$C_r^2 = \frac{\max_t u^2(t)}{\lim_{k\to\infty} \frac{1}{k}\sum_{i=1}^{k} u_i^2(t)} \tag{3.39}$$

A good signal waveform is consequently one that has a small crest factor. The theoretic lower bound of C_r, clearly is 1, which is achieved for binary, symmetric signals. However, the following caution should be mentioned: a binary input will not allow validation against non-linearities. For example, if the true system has a static non-linearity at the input and a binary input is used, the input is still binary after the non-linearity and just corresponds to a scaling. There is consequently no way to detect such a non-linearity using a binary input.

In practice it is suitable to decide upon a frequency band to identify the system in question, and then select a signal with a more or less flat spectrum over this band.

Therefore, in the framework of system identification the basic issue for input signal design can be formulated as: to achieve the desired input spectrum for a signal with a small crest factor as possible [51]. Furthermore, the input should be persistently exciting of a certain order: i.e. that it contains sufficiently distinct frequencies (an open-loop system identification experiment is informative if the input is persistently exciting).

Remark 3.2. *The necessary order of persistent excitation equals the number of parameters to be estimated. If the numerator and denominator of the model have the same number of parameters n, then the input should be persistently exciting of order $2n$. This means that is sufficient to use n sinusoids to identify an nth-order system.*

A simple choice is to let the signal be generated as white Gaussian noise, filtered through a linear filter. The Gaussian signal is theoretically unbounded, so it has to be saturated ("clipped") at certain amplitude. Taking the sign of the signal gives the random binary signal (which can be adjusted to any desired level). The crest factor is thus the ideal 1, the problem is that taking the sign of the filtered Gaussian signal will change its spectrum. Therefore, we don't have full control of shaping the spectrum. The main steps of the previous procedure are shown in Fig. 3.6.

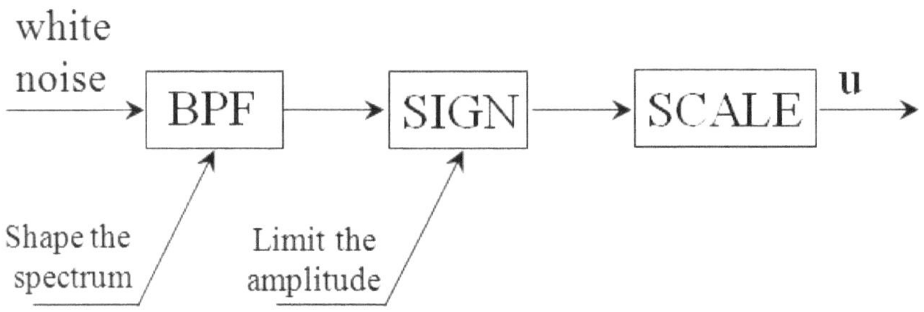

Fig. 3.6 Input signal design - white noise approach.

Another approach is to generate a Pseudo-Random Binary Signal (PRBS), which is deterministic signal with white-noise like properties. The PRBS is generated (shown in Fig. 3.7) by a linear feedback shift register, which consists of n registers z^{-1}.

The output $u(t)$ at the moment t is computed according with equation (3.40):

$$u(t) = modulo\left(\sum_{k=1}^{n} a_k u(t-k), 2\right) \tag{3.40}$$

where a_k are constants and the $modulo(x, 2)$ function is the remainder as x is divided by 2, thus the output only assumes the values 0 and 1.

After u is generated, we can of course change that to any two levels. The vector of past inputs $[u(t-1), ..., u(t-n)]$ can only assume 2^n different values. The sequence u must thus be periodic with a period of at most 2^n. In fact, since n consecutive zeros would make further u's identically zero, we can eliminate that state, and the maximum period of length is $M = 2^n - 1$.

Furthermore, the actual period of the signal depends on the choice of a_k, but it can be shown that for each n exist choices of a_k that give this maximum length. Such choices are shown in Table 3.2 and the corresponding inputs are called maximum length pseudo-random binary signal [51].

The interest in maximum length PRBS follows from the following property: any maximum length PRBS shifting between $\pm\overline{u}$ has the first and second-order properties:

$$\left|\frac{1}{M}\sum_{t=1}^{M} u(t)\right| = \frac{\overline{u}}{M} \tag{3.41}$$

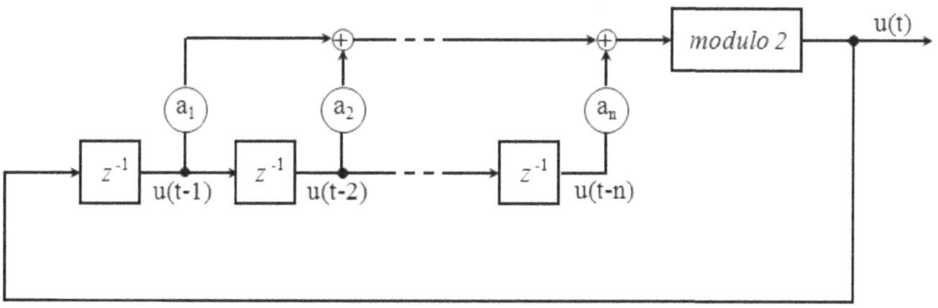

Fig. 3.7 The PRBS generation.

Table 3.2 Maximum length PRBS.

Order n	$M = 2^n - 1$	Sequence - a_k values
2	3	11
3	7	011
4	15	1001
5	31	01001
6	63	100001
7	127	0010001
8	255	11000011
9	511	000100001

$$\phi_{uu}(k) = \frac{1}{M}\sum_{t=1}^{M} u(t)u(t+k) \tag{3.42}$$

which leads to:

$$\phi_{uu}(k) = \begin{cases} \overline{u}^2 & k = 0, \pm M, \pm 2M... \\ -\overline{u}^2/M & otherwise \end{cases} \tag{3.43}$$

where $M = 2^n - 1$ is the maximum length period, as a remark the signal does not have exactly zero mean. The spectrum of the signal is given by equation (3.44) (the terms proportional with $1/M^2$ have been ignored).

$$\Phi_u(\omega) = \frac{2\pi\overline{u}^2}{M}\sum_{k=1}^{M-1} \delta(\omega - 2\pi k/M) \tag{3.44}$$

where $0 \leq \omega \leq 2\pi$ and δ is the Kronecker operator.

In the region $-\pi \leq \omega \leq \pi$ there will be $M - 1$ frequency peaks ($\omega = 0$ excluded). This shows that the maximum length PRBS behaves like "periodic white noise" and is persistently exciting of order $M - 1$. The pseudo-random binary signal and it's power spectral density are shown in Fig. 3.8 and Fig. 3.9, obtained for $n = 9$.

Defining the sample period of the PRBS signal as T_s and decomposing the signal using the Fast Fourier Transform (FFT), the base frequency is defined as: $f_B = 1/MT_s$ and the Nyquist frequency is $f_{Nq} = f_B M/2 = 1/2T_s$. According with the pervious notations the frequency range of the PRBS is:

$$\frac{2\pi}{MT_s} \leq \omega \leq \frac{2\pi}{2T_s} \tag{3.45}$$

Like white random binary noise, PRBS has an optimal crest factor. The advantages and disadvantages of PRBS compared to binary random noise can be summarized as follows:

- as a deterministic signal, PRBS has its second-order properties, when is evaluated over whole periods (for random signals must rely upon the law of large numbers to have good second-order properties for finite samples);
- in order to enjoy the good properties of the PRBS an integer number of periods have to be considered, which limits the choice of experiment length.

As a remark, based on various system identification experiments, when the input signal is a PRBS, which order is higher than 7, has been found that the identified mathematical model is accurate enough in the following frequency range [2]:

$$\frac{2\pi}{100T_s} \leq \omega_{id} \leq \frac{2\pi}{5T_s} \tag{3.46}$$

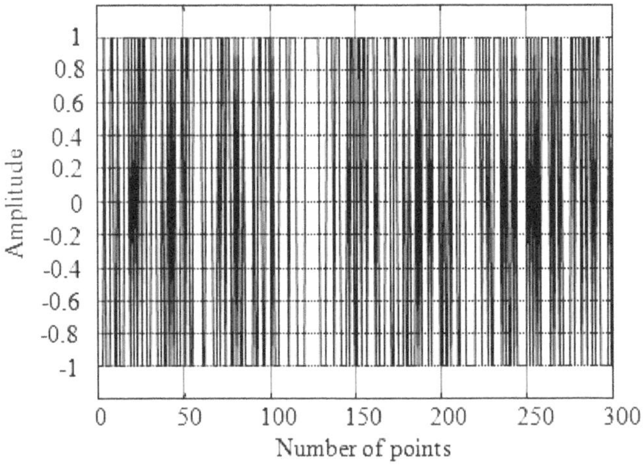

Fig. 3.8 The Pseudo-Random Binary Signal.

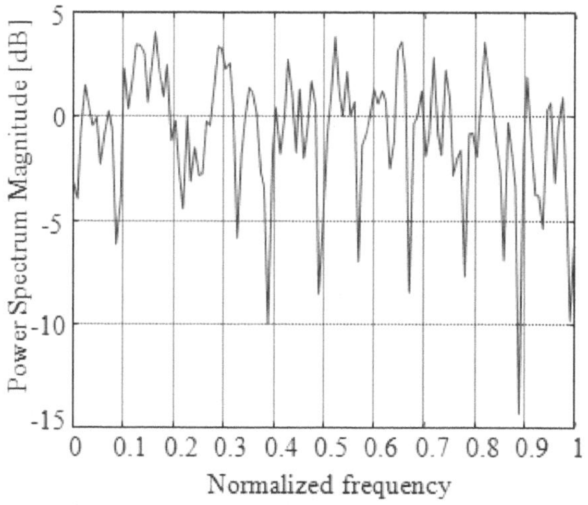

Fig. 3.9 The power spectral density of PRBS.

3.3 Model Validation in Time and Frequency Domain

The parameter estimation procedure picks up the 'best' model within the chosen model structure.

A crucial question then is whether this 'best' model is 'good enough'. The question has several aspects:

- does the model agree sufficiently well with the observed data?
- is the model good enough for our purpose?
- does the model describe the 'true system'?

In practice is recommended to compare the Bode plots of linear models. Comparison between spectral analysis estimates and Bode plots derived from parametric models are especially useful, since they are formed from quite different underlying assumptions.

Residual Analysis

The difference between the measured output and estimated output is called residual:

$$e(t) = y(t) - \hat{y}(t) \tag{3.47}$$

It is clear that these residuals bear information about the quality of the model. A simple and pragmatic starting point is to compute basic statistics for the residuals:

$$S_1 = \max |e(t)| \tag{3.48}$$

$$S_2^2 = \frac{1}{N}\sum_{t=1}^{N} e^2(t) \tag{3.49}$$

The intuitive interpretation would be like this: "The model has never produced a larger residual than S_1 and an average error larger than S_2, for all the data we have seen. It is likely that such a bound will hold also for future data".

The main assumption here is that the residuals do not depend on the particular input used. To check this is reasonable to study the covariance between residuals and past inputs.

$$\phi_{eu}(\tau) = \frac{1}{N}\sum_{t=1}^{N} e(t)u(t-\tau) \tag{3.50}$$

If these numbers are small we have some reason to believe that the measures (3.48) and (3.49) have relevance when the model is applied to other inputs.

Similarly, if we find that the correlation among the residuals themselves:

$$\phi_{ee}(\tau) = \frac{1}{N}\sum_{t=1}^{N} e(t)e(t-\tau) \tag{3.51}$$

is small that is a sign of good model estimate.

When examining the plots of correlation and autocorrelation function we would like to note the followings:

- correlation between $u(t-\tau)$ and $e(t)$ for negative τ is an indication of output feedback in the input, not that the model structure is deficient;
- in the case of ARX models the least squares procedure automatically makes the correlation $\phi_{eu}(\tau) = 0$ for $\tau = 1, \ldots, nb$.

Using the `resid` function of Matlab System Identification Toolbox the autocorrelation function of e and the cross correlation between e and the input u is computed and displayed.

The confidence limits (99%) for these values are also given, based on the hypothesis that the residuals are white and independent of the inputs.

The designed pseudo-random binary signal (PRBS) will be applied for system identification and the main steps of the system identification procedure are illustrated through an example.

Example 3.4. *ARX model identification*

In this case the test system to be identified is defined as:

$$y_n = 0.03y_{n-1} - 0.01y_{n-2} + 3u_n - 0.5u_{n-1} + 0.2u_{n-2} \tag{3.52}$$

It is left to the reader to choose the order of pseudo-random binary signal such that the system identification experiment is informative (the input signal is persistently exciting). Furthermore, to make the identification experiment more realistic the output is corrupted by white noise e.g. $N(0, 0.5^2)$.

The input and output data are shown in Fig. 3.10. The main steps of the identification are illustrated in the Matlab session 3.3.1.

The estimated output is shown in Fig. 3.11 by the dashed line and the solid line marks the real system output.

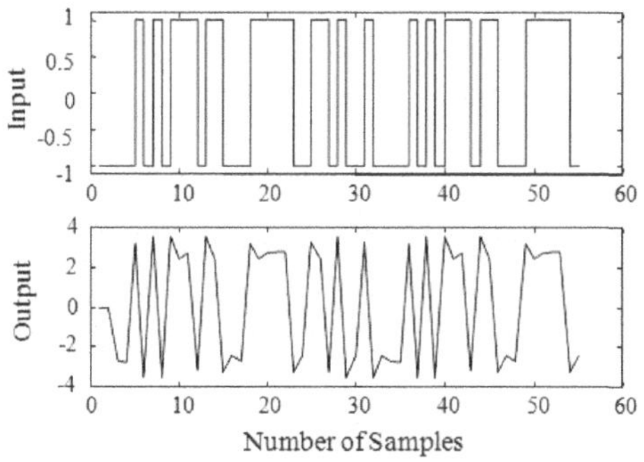

Fig. 3.10 Input-output data - ARX model.

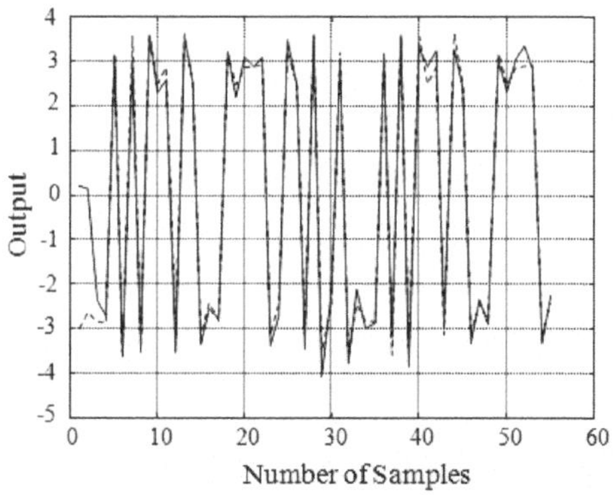

Fig. 3.11 Measured and estimated output.

MATLAB session 3.3.1: Identification of the ARX model.

```
% load the input-output data - it is left to the user to
generate the data pairs
>> load dataarx.mat;
% where u - is the input data, yn - is the output data
affected by white noise
% Using input and output data pairs - identify the model
using LMS method

% select an ARX model structure
>> na = 2; nb =3; nk =0;
% with na A parameters
% with nb B parameters
% with nk delays
>> th = arx([yn'u'],[nb nf nk]);

% COMPARE THE MODELS - SYSTEM OUTPUTS
>> plot(yn)
>> pause
>> hold on
>> grid
>> ylms = idsim(u',th);
>> plot(ylms,'g')
>> pause
>> close

% Plot the impulse response of the estimated model
>> xx = [1 zeros(1,5)];
>> tt = 0:5;
% model identified using the LMS
>> yylms = idsim(xx',th);
>> plot(tt,yylms,'r')
>> pause

% Plot the frequency response
>> gg=th2ff(th);
>> bodeplot(gg)
```

Fig. 3.12 shows a comparison between the true impulse response (solid line) and estimated one (dotted line). The frequency response of the derived ARX model is illustrated in Fig. 3.13.

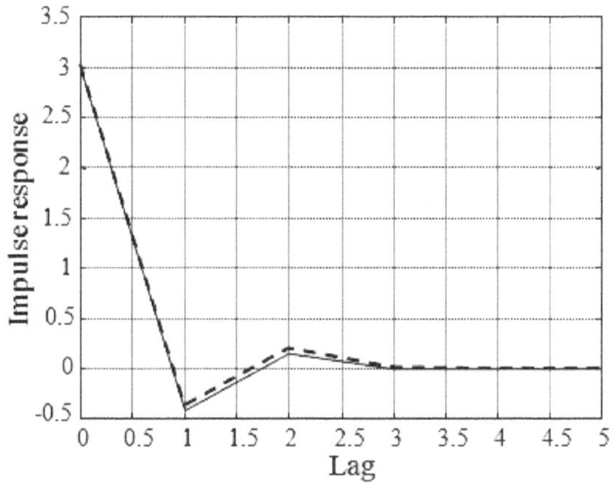

Fig. 3.12 Impulse response - ARX model.

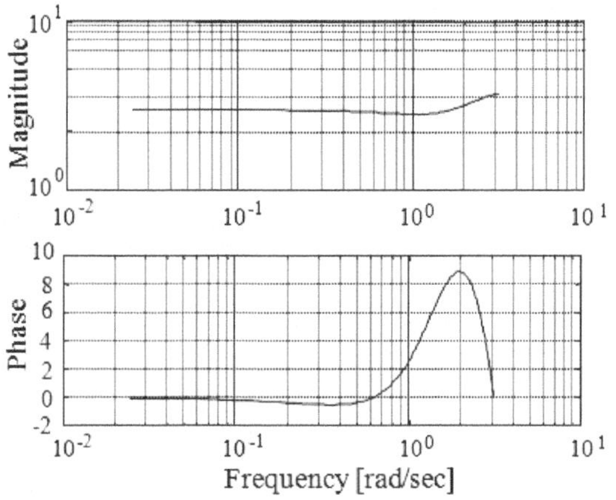

Fig. 3.13 Frequency response - ARX model.

The model is validated in the time domain (using correlation analysis) as well as in the frequency domain. The model validation procedure is illustrated in the Matlab session 3.3.2 and the main results are presented in Fig. 3.14 and Fig. 3.15.

MATLAB session 3.3.2: Validation of ARX model.

```
% MODEL VALIDATION
% compute residuals
% correlation and autocorrelation functions
>> resid(z,th);
>> pause
>> close

% compute the frequency function
>> gg=th2ff(th);
% Estimate the frequency response by spectral analysis
>> gsa = spa(z);
% Compare the Bode plots
>> bodeplot([gg gsa])
>> pause
```

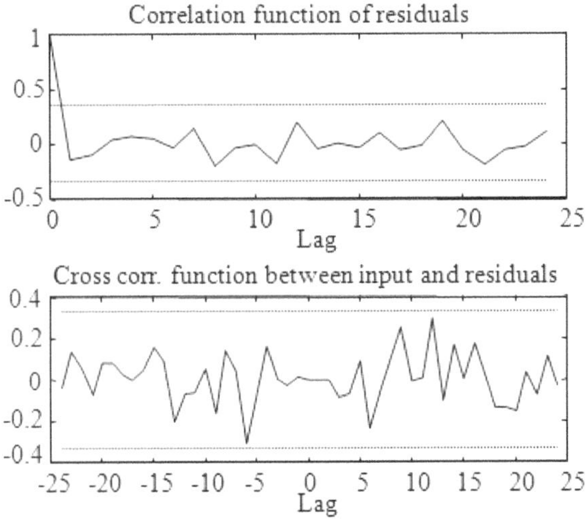

Fig. 3.14 Model validation - correlation analysis.

Finally, is left for the reader to design an off-line system identification experiment in order to identify the system's parameters, given in the example below.

Example 3.5. *Consider the plant having the transfer function:*

$$P(s) = \frac{1}{sL + R} \tag{3.53}$$

with parameters $R = 6.8\Omega$ *and* $L = 0.5H$.

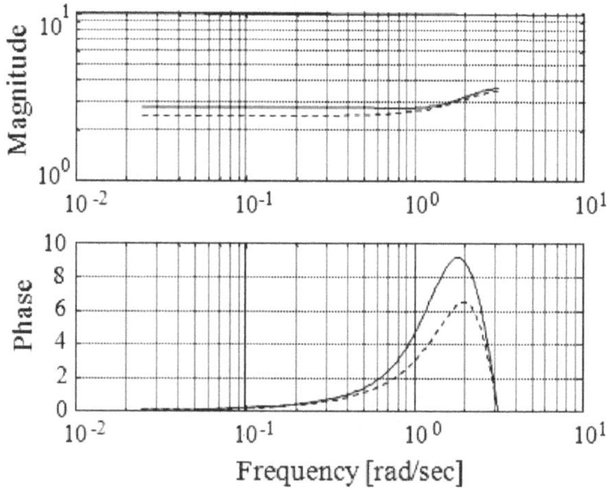

Fig. 3.15 Model validation - frequency domain.

White noise shall be added to the output signal to see the influence of noise on parameter estimation.

3.4 Model-Order Reduction Methods

The system identification procedure usually leads to a high-order mathematical model, which is not suitable for control system design purposes. Therefore model-order reduction is required. In a classical control framework, this is usually done by using so-called pole-zero cancellation. This section is going to discuss model-order reduction methods, based on balanced truncation technique [27].

We will need some mathematical background to understand the idea behind model-order reduction and how to use the modern software tools to perform this task.

3.4.1 Model-Order Reduction

In order to introduce this topic we will start with some useful results from matrix theory.

The Lyapunov equations and inequalities come in the two dual forms:

$$\begin{aligned} A^*X + XA + Q = 0 \\ AX + XA^* + Q = 0 \end{aligned} \tag{3.54}$$

In both cases A, Q are given square matrices, and X is the unknown, where * denotes the adjoint of a matrix or operator.

Here, we briefly recall that for a matrix $A \in C^{n \times m}$:

$$A = \begin{bmatrix} a_{11} & \dots & a_{1n} \\ . & \dots & . \\ a_{m1} & \dots & a_{mn} \end{bmatrix} \tag{3.55}$$

the transpose is:

$$A^T = \begin{bmatrix} a_{11} & \dots & a_{m1} \\ \vdots & \vdots & \vdots \\ a_{1n} & \dots & a_{nm} \end{bmatrix} \tag{3.56}$$

and the adjoint is defined as:

$$A^* = \begin{bmatrix} a_{11}^* & \dots & a_{m1}^* \\ . & \dots & . \\ a_{1n}^* & \dots & a_{nm}^* \end{bmatrix} \tag{3.57}$$

where a_{ij}^* is the complex conjugate of a_{ij}.

Theorem 3.2. *Suppose A and Q are square matrices and A is Hurwitz, then:*

$$X = \int_0^\infty e^{A^*\tau} Q e^{A\tau} d\tau \tag{3.58}$$

is the unique solution to the Lyapunov equation

Theorem 3.3. *The matrix A is Hurwitz iff there exist $X > 0$*

$$A^* X + XA < 0 \tag{3.59}$$

(this is one way to verify if a system is stable or not).

Observability Operator and Gramiam

Let us focus on the autonomous system given by:

$$\begin{aligned} \dot{x}(t) &= Ax(t),\ x(0) = x_0 \in C^n \\ y(t) &= Cx(t) \end{aligned} \tag{3.60}$$

where A is a Hurwitz matrix.

One way to check the observability of a system is to check the rank of the following matrix:

$$Q^* = [C^* \ A^*C^* \ \dots \ (A^*)^{n-1}C^*] \tag{3.61}$$

whether the matrix has rank n.

The above rank test can only provide a *yes* or *no* answer related to the observability of the system. We need to measure the, "observability" of each state, in this aim we introduce the concept of **gramiam**.

In line with this reasoning, it seems natural that the gramiams could be used as a basis of a model reduction scheme, based on the elimination of weekly observable states. We will return to this idea later, but first we study observability and controllability gramiams.

The Lyapunov solution denoted by Y_0 satisfying:

$$A^*Y_0 + Y_0A + C^*C = 0 \tag{3.62}$$

is called the **observability gramian** of (C, A), which is:

$$Y_0 = \int_0^{\infty} e^{A^*\tau} C^* C e^{A\tau} d\tau \tag{3.63}$$

The gramiam measures "how observable" a given initial condition is.

Let $\eta_1 \geq \eta_2 \geq ... \geq \eta_n \geq 0$ to be the eigenvalues of $\sqrt{Y_0}$ and $v_1, ..., v_n$ their respective unit-norm eigenvectors. Then the v_k give the directions of the principal axes of the ellipsoid and η_k the length of each axis.

If $\eta_k = 0$ for some k, the corresponding v_k produces no output energy and is therefore unobservable. If η_k, η_l are both nonzero but, for example $\eta_k \geq \eta_l$ then the output energy resulting from initial state v_l is much smaller than that observed when the initial condition is v_k.

Thus both states are observable, but intuitively state v_k is "more observable" than state v_l. We conclude that the observability gramiam provides us with a way to assess the relative observability of various directions in the state space.

Since Y_0 is positive semidefinite this set is and ellipsoid and is called the **observability ellipsoid**.

Controllability Gramiam

Given the state equation $\dot{x}(t) = Ax(t) + Bu(t)$ we call the matrix

$$U_c = [B \; AB \; A^2B \; ... \; A^{n-1}B] \tag{3.64}$$

the associated controllability matrix.

One way to check the controllability of the system is to check the rank of the previous matrix. If the matrix rank is n that means all states are controllable. The above rank test can only provide a *yes* or *no* answer related to the controllability of the system.

The **controllability grammian** is defined in a similar way and is denoted by X_c and satisfies the Lyapunov equation:

$$AX_c + X_cA^* + BB^* = 0 \tag{3.65}$$

dual to the observability case

$$X_c = \int_0^{\infty} e^{A\tau} BB^* e^{A^*\tau} d\tau \tag{3.66}$$

Let $\mu_1 \geq \mu_2 \geq ... \geq \mu_n \geq 0$ to be the eigenvalues of $\sqrt{X_c}$ and $v_1, ..., v_n$ their corresponding orthonormal eigenvectors. We can conclude that if $\mu_k = 0$ then v_k is an unreachable state. In the same way if $\mu_k \geq \mu_l$ then the state direction v_k is "more" controllable than direction v_l.

The preceding sections have given us geometric ways, in terms of the grammians introduced, to assess which directions in the state space are strongly or weakly controllable and observable.

Balanced Realization

We concluded that states corresponding to small eigenvalues of the observability gramiam are not very observable. Now a natural question arises: does this mean that such states do not contribute much to the input-output behaviour of the above system? The answer is not necessarily, since such states may be very controllable (see Fig. 3.16).

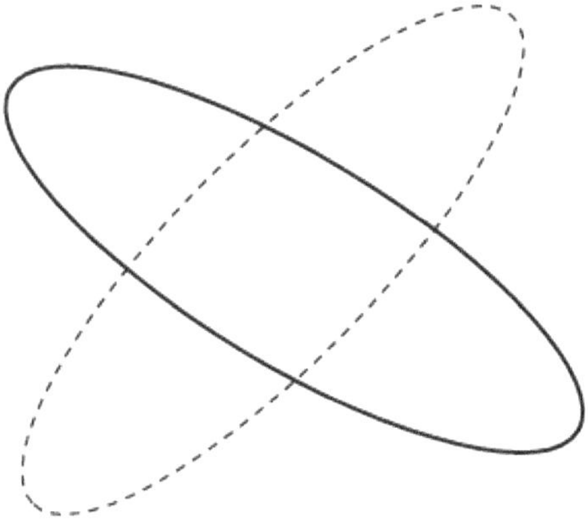

Fig. 3.16 Unbalanced system ellipsoids.

Therefore, we reason that the most intuition would be gained about the system if the controllability and observability ellipsoids were exactly aligned.

A state space realization such that the controllability and observability gramians are equal and diagonal, is called **balanced realization**.

Theorem 3.4. *Suppose (A, B, C) is a controllable and observable realization. Then there exists a state transformation T such that the equivalent realization:*

$$(\tilde{A}, \tilde{B}, \tilde{C}) = (TAT^{-1}, TB, CT^{-1}) \tag{3.67}$$

satisfies

$$\tilde{X_c} = \tilde{Y_o} = \sum \tag{3.68}$$

where $\sum > 0$ and is diagonal.

Remark 3.3. *In analogy to the matrix case, the square roots of the eigenvalues of $\tilde{Y}_o\tilde{X}_c$ are called the* **Hankel singular values** *of the system.*

The Hankel operator plays an important part in systems theory, as well as pure operator theory, and is intimately related to the question of approximation the ∞-norm.

Theorem 3.5. *Consider the following system $P(s) = C(sI - A)^{-1}B$, where A is a Hurwitz matrix. Then*

$$\|P\|_\infty \leq 2(\sigma_1 + \cdots + \sigma_n) \tag{3.69}$$

where the σ_k are the Hankel singular values of P.

Model-Order Reduction Based on Balanced Truncation

Frequently in modelling or control synthesis we would like to use a simple model that approximates a more complex one.

In our context, complexity is measured by the dimensionality of state-space realizations. Given a system, we would like to reduce the order of a state-space realization, while keeping the system input-output properties approximately the same.

An extreme case of model reduction we have already studied is the elimination of uncontrollable or unobservable states. This can be done without affecting the input-output transfer function.

We are now ready to take the next step, that is, to allow some error, and study the following problem: given a transfer function $P(s)$ of order n, find a transfer function of order $P_r(s)$ of order $r \leq n$, such that P and P_r are close.

Before this problem has a precise meaning, a notion of error between transfer functions must be chosen; indeed there are many metrics that can be used, we will focus on the $\mathcal{H}_\infty$ model reduction problem.

Given a transfer function

$$P(s) = \left[\begin{array}{c|c} A & B \\ \hline C & D \end{array}\right](s) \tag{3.70}$$

where $A \in \Re^{n\times n}$ is Hurwitz, find a lower-order function:

$$P_r(s) = \left[\begin{array}{c|c} A_r & B_r \\ \hline C_r & D_r \end{array}\right](s) \tag{3.71}$$

where $A_r \in \Re^{r\times r}$ and is Hurwitz, such that: $\|P - P_r\|_\infty$ is minimized. Here we remark that if $D = 0$ means that the transfer function is strictly proper.

As stated in [27], $\mathcal{H}_\infty$ model reduction problem is open, in the sense that there is no known computationally tractable method to obtain the optimal approximation of a given order.

Despite that let's focus on the balanced truncation method. It is possible to show that a balanced realization for a transfer function can always be found. Formally, this means a realization such that the gramiams satisfy:

$$X_c = Y_o = \begin{bmatrix} \sigma_1 & & & \\ & \sigma_2 & & \\ & & \cdots & \\ & & & \sigma_n \end{bmatrix} \tag{3.72}$$

where $\sigma_1, ..., \sigma_n$ are the Hankel singular values

Also we remark that if a given σ_k was small compared with the other eigenvalues, the corresponding state had the interpretation of being weakly controllable and observable. The technique for model reduction we now consider, essentially amounts to discharging or truncating such states.

Here we necessarily assume that the realization (A, B, C) is a balanced realization and we order the Hankel singular values:

$$\sigma_1 \geq \sigma_2 \geq ... \geq \sigma_n \tag{3.73}$$

Let us suppose we want to find a realization of order $r < n$. We assume that the strict inequality $\sigma_{r+1} < \sigma_r$ is satisfied. We only attempt to truncate the system at an order where the Hankel singular values have clear separation.

Then we compatibly partition our realization as:

$$A = \begin{bmatrix} A_{11} & A_{12} \\ A_{21} & A_{22} \end{bmatrix} \quad B = \begin{bmatrix} B_1 \\ B_2 \end{bmatrix} \quad C = \begin{bmatrix} C_1 & C_2 \end{bmatrix} \tag{3.74}$$

where $A_{11} \in \Re^{r \times r}$. The reduced-order model is chosen to be:

$$P_r(s) = \left[\begin{array}{c|c} A_{11} & B_1 \\ \hline C_1 & 0 \end{array} \right] (s) \tag{3.75}$$

The minimum requirement on P_r is that A_{11} be Hurwitz, and this is indeed the case.

Lemma 3.2. *Bound for the balanced truncation error. Suppose the Hankel singular values of P satisfy:*

$$\sigma_1 \geq \sigma_2 \geq ... \geq \sigma_r > \sigma_{r+1} = \sigma_{r+2} = ... = \sigma_n \tag{3.76}$$

If P_r is obtained by balanced truncation, then:

$$\|P - P_r\|_\infty \leq 2\sigma_{r+1} \tag{3.77}$$

In this section we described a method by which we reduced the dynamic size of the state-space model, with the ∞-norm as the quality measure on the error. We conclude this section with a remark.

Remark 3.4. *The problem of minimizing the $\|P - P_r\|_\infty$ remains computationally difficult and the balanced truncation method is not optimal.*

We can solve the model reduction problem and we know that the reduction leads to a certain error. The difficulty is that the above conditions, while easy to state, are not computationally tractable [27].

The model-order reduction is illustrated in the framework of Matlab session 3.4.1.

MATLAB session 3.4.1: Model-order reduction.

```
% model transformation - to state space representation
>> [A, B, C, D] = th2ss(th);

% ***** model reduction *************
% start with a correctly normalized (balanced) model
>> [ab, bb, cb, g] = dbalreal(A, B, C);

% now eliminates those states whose gramian values are
below a threshold - for example g(1)/50
>> elim = find(g < g(1)/50);
>> [ar, br, cr, dr] = dmodred(ab, bb, cb, D, elim);
```

Now we reached the point, where we can say that we can derive:

- a high-order mathematical model, which approximates well (in a defined frequency range) the linear plant, however is not suitable for controller design
- a low-order mathematical model derived from the high-order mathematical model (using model-order reduction), which is suitable for controller design.

The difference between the high-order and low-order model can be viewed as unstructured uncertainty.

In addition to the unstructured uncertainty we might have the situation that the parameters of the low-order mathematical model are not known exactly, then we have parametric uncertainty as well.

We will illustrate these uncertainties in the next section.

3.5 Nominal Plant and Plant Uncertainties

Let us start with a simple structured parametric uncertainty and show how is modeled.

Example 3.6. *Structured parameter uncertainty*

Let us assume that the parameter a is changing between two extreme values: $a_{min} \leq a \leq a_{max}$. Moreover, let us note the mean value with $a_{med} = (a_{max} -$

$a_{min})/2$ and the upper bound of parameter variation with: $a_{vmax} = a_{max} - a_{med}$, then the parameter a can be expressed as:

$$a = a_{med} + \Delta a_{vmax}, \ \Delta \in [-1, 1] \tag{3.78}$$

For $\Delta = 0$ we obtain $a = a_{med}$, which can be considered as nominal value in the design process.

The parameter uncertainty model described by equation (3.78) is represented in Fig 3.17.

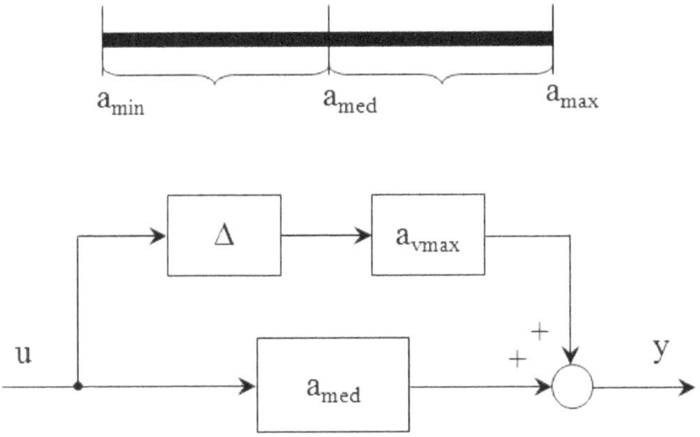

Fig. 3.17 Parameter uncertainty model.

Next, let us consider the plant belonging to a structured set:

$$\mathcal{P} = \left\{ \frac{1}{s^2 + as + 1} : a_{min} \leq a \leq a_{max} \right\} \tag{3.79}$$

With the previously introduced notations we can write:

$$P(s) = \frac{1}{s^2 + a_{med}s + 1} \tag{3.80}$$

and

$$W(s) = a_{vmax}s \tag{3.81}$$

Then

$$\mathcal{P} = \frac{P(s)}{1 + \Delta W(s)P(s)} : -1 \leq \Delta \leq 1 \tag{3.82}$$

Next, let us illustrate the concept of unstructured parametric uncertainty considering another example.

Example 3.7. *Let us consider a flexible beam attached to a rigid body (for example a large antenna attached to a space satellite). The input is the torque* *applied to the beam by the rigid body and the output is the displacement at the other end of the beam. We assume that the magnitude of the torque is small, so that the displacements are small. Therefore, the system can be linearized around the equilibrium position.*

Considering a linearized model, the input/output behavior of this system is given by the transfer function:

$$P_\infty(s) = \frac{k_0}{s^2} + \sum_{k=1}^{\infty} \frac{\alpha_k \omega_k^2}{s^2 + 2\xi_k \omega_k s + \omega_k^2} \tag{3.83}$$

where ω_k's represent the frequency of natural oscillations (modes of the system), ξ_k's are the corresponding damping ratios for each mode and α_ks are the weights of each mode. Typically, $\alpha_k \to 0$ as $k \to \infty$. When α_k's converge to zero very fast, we can truncate the higher modes and obtain a finite dimensional model for the plant.

To illustrate this point, let us define the approximate plant transfer function as:

$$P_N(s) = \frac{k_0}{s^2} + \sum_{k=1}^{N} \frac{\alpha_k \omega_k^2}{s^2 + 2\xi_k \omega_k s + \omega_k^2} \tag{3.84}$$

The approximation error between the 'true' plant transfer function and the approximate model is:

$$\Delta P_N = P_\infty(s) - P_N(s) \tag{3.85}$$

As conclusion, for a plant a lower-order model denoted by $P_N(s)$ can be found. Furthermore, an upper bound of the worst approximation error can be obtained.

$$|P_\infty(j\omega) - P_N(j\omega)| < \delta_N(\omega)\ \forall \omega \tag{3.86}$$

Next, let us consider the following model:

$$P_1(s) = \frac{k_0}{s^2} + \frac{\alpha_1 \omega_1^2}{s^2 + 2\xi_1 \omega_1 s + \omega_1^2} \tag{3.87}$$

Suppose that an upper bound $\delta_1(\omega)$ is determined as above for model uncertainty.

In practice, the parameters of the model $P_1(s)$ are determined from the physical parameters of the beam (such as inertia and elasticity). If these parameters are not known exactly, then we have parametric uncertainty as well.

Let's say that the parametric uncertainty bound is denoted by $\Delta_1(\omega)$:

$$|P_1(j\omega) - P_{1n}(j\omega)| < \Delta_1(\omega)\ \forall \omega \tag{3.88}$$

Parametric uncertainty bound and the bound for unmodeled dynamic uncertainty can be combined to get an upper bound the overall plant uncertainty.

$$|\Delta_a(j\omega)| > |\delta_1(j\omega)| + |\Delta_1(j\omega)| \; \forall\omega \tag{3.89}$$

We already emphasized that no mathematical system can exactly model a physical system. For this reason we must be aware of how modeling errors might adversely affect the performance of a control system.

The basic technique is to model the plant as belonging to a set $\mathcal{P}$. Such a set can be either structured or unstructured.

The plant can belong to structured or unstructured sets, however unstructured sets are more important, for two reasons:

- all models used in feedback design should include some unstructured uncertainty to cover unmodelled dynamics particularly at high frequency
- for a specific type of unstructured uncertainty it's possible to develop simple, general analysis methods.

Two frequently used uncertainty models are described next.

3.5.1 Multiplicative Uncertainty Model

Suppose that the nominal transfer function is P and consider that the perturbed plant transfer function has the form (shown in Fig. 3.18):

$$\tilde{P} = (1 + \Delta W_T)P \tag{3.90}$$

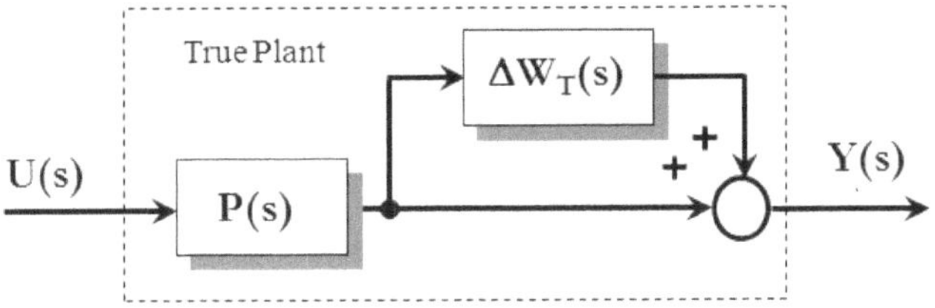

Fig. 3.18 Multiplicative uncertainty.

Here W_T is a fixed stable transfer function (the weight) and Δ is a variable stable transfer function satisfying $\|\Delta\|_\infty \leq 1$.

The idea behind this uncertainty model is that:

$$\Delta W_T = \frac{\tilde{P} - P}{P} \tag{3.91}$$

In practice W_T is chosen to be the upper bound of Δ_m.

$$\Delta_m = \frac{\tilde{P} - P}{P} \tag{3.92}$$

Hence if $\|\Delta\|_\infty \leq 1$ then:

$$|\Delta_m(j\omega)| = \left|\frac{\tilde{P}(j\omega)}{P(j\omega)} - 1\right| \leq | W_T(j\omega) |, \forall\omega \tag{3.93}$$

So $|W_T(j\omega)|$ provides the uncertainty profile. Typically, $|W_T(j\omega)|$ is a (roughly) increasing function of ω, uncertainty increasing with increasing frequency.

Remark 3.5. *The multiplicative uncertainty model is not suitable for every application because the uncertainty set is sometime too coarse approximation. Other uncertainty models are possible besides multiplicative perturbations.*

3.5.2 Additive Uncertainty Model

The additive uncertainty model is represented in Fig. 3.19, where W_Q is a fixed stable transfer function (the weight) and Δ is a variable stable transfer function satisfying $\|\Delta\|_\infty \leq 1$.

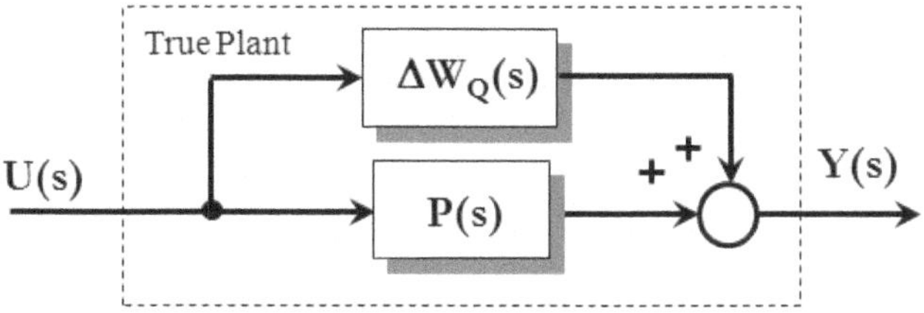

Fig. 3.19 Additive uncertainty.

$$\Delta W_Q = \tilde{P} - P \tag{3.94}$$

In practice W_Q is chosen to be the upper bound of Δ_a.

$$\Delta_a = \tilde{P} - P \tag{3.95}$$

$$|\Delta_a(j\omega)| = |\tilde{P}(j\omega) - P(j\omega)| \leq |W_Q(j\omega)|, \forall\omega \tag{3.96}$$

If a multiplicative uncertainty bound $W_T(s)$ is given instead of the additive uncertainty bound $W_Q(s)$, then we use the following relationship:

$$\tilde{P} = P + \Delta_a = P(1 + P^{-1}\Delta_a) \tag{3.97}$$

thus

$$|P^{-1}(j\omega)\Delta_a(j\omega)| \leq |W_T(j\omega)|, \forall\omega \tag{3.98}$$

Example 3.8. *Multiplicative uncertainty model for a plant with uncertain parameters.*

Consider the plant $\tilde{P}(s)$ with uncertain parameters L and R:

$$\tilde{P}(s) = \frac{1}{sL + R} \tag{3.99}$$

where $L \in [0.4, 0.6]H$ and $R \in [6.65, 6.95]\Omega$.

Derive a multiplicative uncertainty model assuming that the nominal model is given by:

$$P(s) = \frac{1}{sL_{med} + R_{med}} \tag{3.100}$$

with $L_{med} = 0.5H$ and $R_{med} = 6.8\Omega$.

The multiplicative uncertainty model is given by:

$$\Delta_m = \frac{\tilde{P} - P}{P} = \frac{s(L_{med} - L) + R_{med} - R}{sL + R} \tag{3.101}$$

and is represented in Fig. 3.20 for different values of L and R.

Moreover, the weighting function $W_T(s)$ is made larger than the multiplicative uncertainty $\Delta_m(s)$ at any frequency:

$$|\Delta_m(j\omega)| \leq |W_T(j\omega)|, \forall\omega \tag{3.102}$$

with $W_T(s)$ given by:

$$W_T(s) = \frac{s + 1}{2(s + 20)} \tag{3.103}$$

and represented in Fig. 3.20 by the dashed line.

Since the set $\Delta W_T(s)$ with $\|\Delta\|_\infty \leq 1$ is larger than the actual set of interest $\Delta_m(s)$, which means that if a controller achieves design objectives for all plants in the set $P(s)(1 + \Delta W_T(s))$, then it is guaranteed to work for all plants in $\tilde{P}(s) = P(s)(1 + \Delta_m(s))$.

Example 3.9. *Additive uncertainty model for a plant with uncertain time delay.*

Consider the plant $\tilde{P}(s)$ with uncertain time delay T_d:

$$\tilde{P}(s) = \frac{1}{(s + 1)(s + 2)} e^{-sT_d} \tag{3.104}$$

where $T_d \in [0, 0.1]$.

Derive an additive uncertainty model assuming that the nominal model is given by:

$$P(s) = \frac{1}{(s + 1)(s + 2)} \tag{3.105}$$

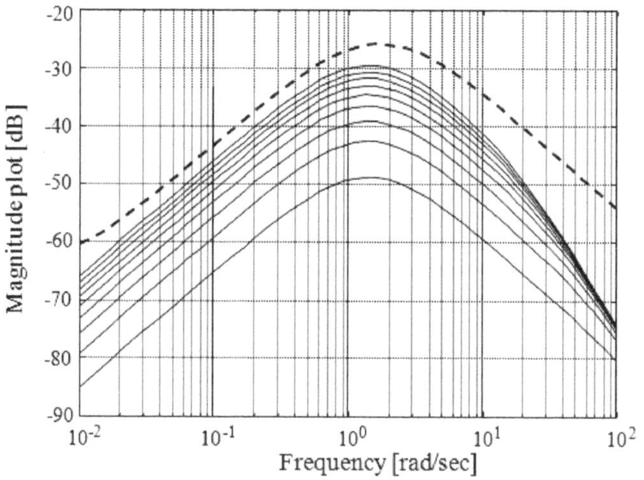

Fig. 3.20 Example - multiplicative uncertainty.

The additive uncertainty model is given by:

$$\Delta_a = \tilde{P} - P = P(e^{-sT_d} - 1) \tag{3.106}$$

and is represented in Fig. 3.21 for different values of time delay $T_d \in [0, 0.1]$.

Moreover, the weighting function $W_Q(s)$ is made larger than the additive uncertainty $\Delta_a(s)$ at any frequency:

$$|\Delta_a(j\omega)| \leq |W_Q(j\omega)|, \forall\omega \tag{3.107}$$

with $W_Q(s)$ given by:

$$W_Q(s) = \frac{100s + 1}{500(s + 1)(s + 3)} \tag{3.108}$$

and represented in Fig. 3.21 by the dashed line.

Since the set $\Delta W_Q(s)$ with $\|\Delta\|_\infty \leq 1$ is larger than the actual set of interest $\Delta_a(s)$, which means that if a controller achieves design objectives for all plants in the set $P(s) + \Delta W_Q(s)$, then it is guaranteed to work for all plants in $\tilde{P}(s) = P(s) + \Delta_a(s)$.

Definition 3.2. *The available bandwidth is the frequency range over, which the unstructured multiplicative uncertainty is substantially less than the unity. The available bandwidth is noted by ω_a.*

Remark 3.6. *Let us note that the available bandwidth is not a function of the compensator or of the control design process, rather it is an a priori constraint imposed by the physical hardware, which we use in the control loop. Most important the available bandwidth is always finite.*

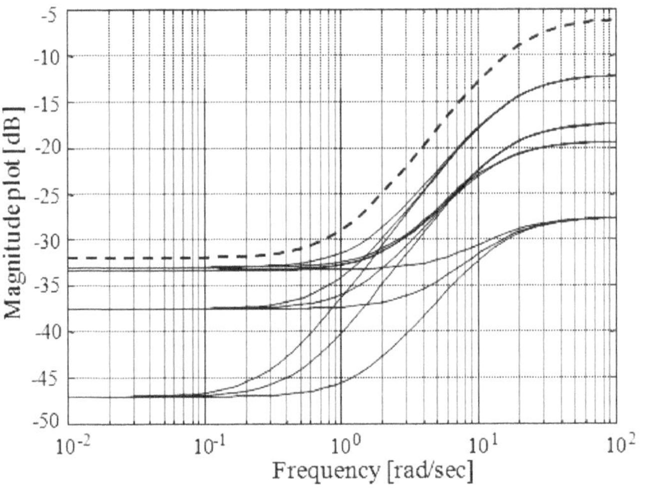

Fig. 3.21 Example - additive uncertainty.

3.6 Practical Examples – System Identification

3.6.1 Brushless d.c. Drive's Identification

Let us consider a vector controlled brushless d.c. drive as shown in Fig. (3.22). The final goal is to design a controller for this drive system. However, at this stage let us focus only on model building. Using system identification techniques, we would like to derive a nominal plant and an upper bound for the plant uncertainty, which will be used for controller design.

The first step is to design the system identification experiment. We will choose as input signal the i_q^* reference current and as output the motor speed. In order to persistently excite the system, we choose a 7-th order pseudo-random binary signal (PRBS) as input signal. The persistence excitation of the chosen PRBS is much higher than the order of physical model.

If we assume that the inverter and the sensors are ideal the simplified physical model will lead to second-order model. This physical insight into the system suggests that we might try to derive a second-order nominal plant and a related plant uncertainty.

Defining a frequency range for identification as ω_{id} the sampling time T_s is selected according to equation (3.109):

$$\frac{2\pi}{100T_s} < \omega_{id} < \frac{2\pi}{5T_s} \tag{3.109}$$

In our case $T_s = 0.001s$, that means $50[rad/sec] < \omega_{id} < 1250[rad/sec]$.

The next step is to collect the input/output data, this is shown in Fig. 3.23.

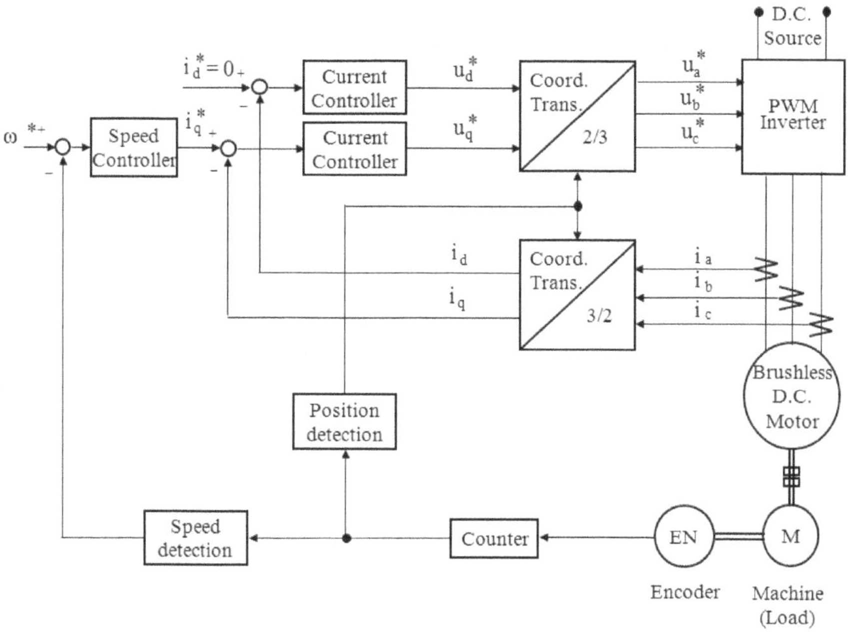

Fig. 3.22 Vector controlled brushless d.c. drive.

The collected data is de-trended and outliers are eliminated. In order to get a good fit, we select an ARX model structure and a model having 15th-order. The selected model-order is too high for controller design purposes but gives a good fit between measured and estimated data. This high-order model is validated using residual analysis, as we have shown in this chapter.

The next step is model-order reduction based on balanced truncation technique. We truncate the high-order model to a 2nd-order model, according to a simplified physical model.

A comparison between the identified mathematical model and the reduced-order model is presented in Fig. 3.24 (gain plot), where the dotted line corresponds to the high-order model (15th-order model) and the solid line to the reduced-order model (2nd-order model).

The transfer function of the reduced-order model obtained using the balanced truncation technique can be written as:

$$P(s) = \frac{K_{gain}}{(s+\omega_1)(s+\omega_2)} \tag{3.110}$$

where: $K_{gain} = 1/20[\text{rad/sec/A}]$, $\omega_1 = 2.5[\text{rad/sec}]$ and $\omega_2 = 800[\text{rad/sec}]$.

The second-order model can approximate well the high-order model in the frequency range from 50[rad/sec] till 1000[rad/sec].

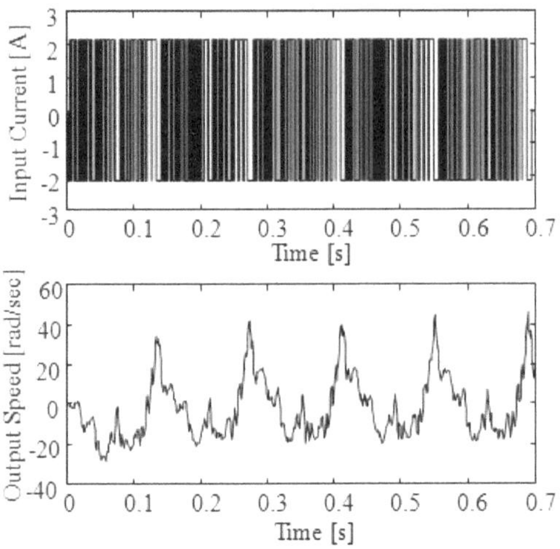

Fig. 3.23 BLDC drive identification.

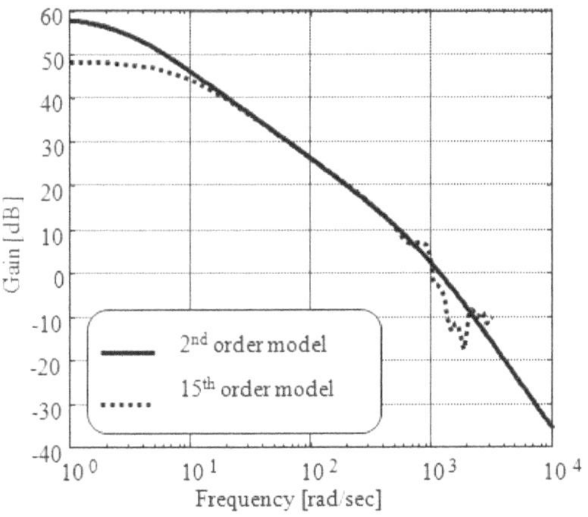

Fig. 3.24 BLDC drive Bode diagram.

The last step is to derive the plant uncertainty. As we mentioned all models used in feedback design should include some unstructured uncertainty to cover un-modeled dynamics, particularly at high frequency.

The plant multiplicative uncertainty is:

$$\Delta_m = \frac{\widetilde{P(s)} - P(s)}{P(s)} \tag{3.111}$$

where $P(s)$ is the nominal plant (2nd-order model) and $\widetilde{P(s)}$ is the real plant model (15th-order model).

Then the weighting function $W_T(s)$ is made larger than the multiplicative uncertainty $\Delta_m(s)$ at any frequency [25].

$$|\Delta_m(j\omega)| \leq |W_T(j\omega)| \; \forall\omega \tag{3.112}$$

Anticipating a little bit - in the framework of Matlab's Robust Control Toolbox - it is required that the $W_T(s) \cdot P(s)$ transfer function must be proper, i.e., bounded as $s \to \infty$. However, $W_T(s)$ may be improper, the possibly improper transfer function $W_T(s)$, which satisfies the previous inequality might be chosen as (shown in Fig. 3.25, dotted line):

$$W_T(s) = 0.1\frac{s^2 + 400s + 200^2}{200^2} \tag{3.113}$$

The framework of Matlab's Linear Matrix Inequalities Toolbox does not allow improper transfer functions for $W_T(s)$, then it can be approximated over any desired frequency range by a proper one - the previous weighting function is rewritten for instance as:

$$W_T(s) = 0.1\frac{s^2 + 400s + 200^2}{200^2}\frac{2000^2}{s^2 + 2800s + 2000^2} \tag{3.114}$$

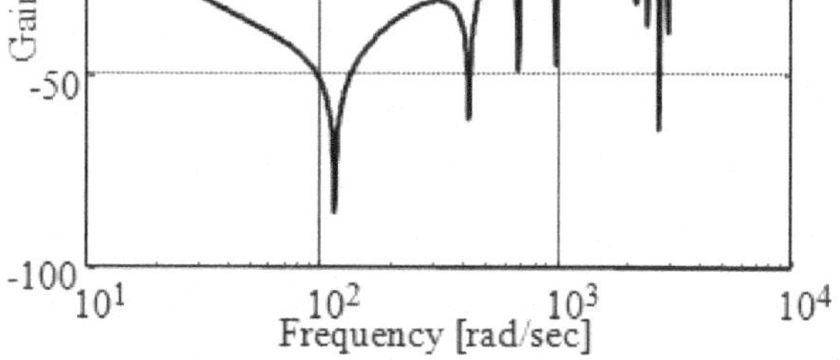

Fig. 3.25 Plant uncertainty - BLDC drive.

3.6.2 Identification of a Fuel Cell

The next example, which we present here is the system identification of a Proton Exchange Membrane (PEM) fuel cell. The purpose is cell parameters' identification for state diagnosis. At this stage we do not have the purpose to control the fuel cell, therefore we will derive only a parametric model.

In order not to complicate things, we will view the fuel cell as battery cell. The equivalent circuit model is shown in Fig. 3.26 and the static characteristics is presented in Fig. 3.27.

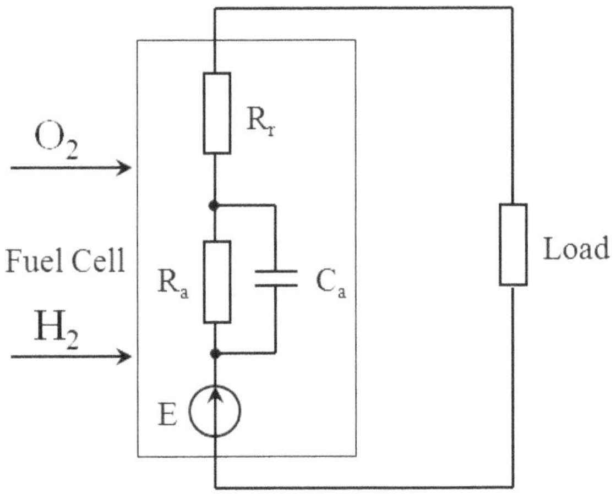

Fig. 3.26 Equivalent circuit model.

The resistor R_r models the ohmic losses. A change in current gives an immediate change in the voltage drop across this resistor. The resistor R_a models the so-called "activation overvoltge", and the capacitor C_a "smoothes" any voltage drop across this resistor [31].

Based on this simplified circuit model the internal impedance of the fuel cell can be written as:

$$Z(s) = R_r + \frac{R_a}{sR_aC_a + 1} \tag{3.115}$$

In order to identify the parameters we have to choose at least two different operating points on the linear region of the characteristics (we would do the same if we would like to find out the internal resistance).

Around each operating point we persistently excite the system with a pseudo-random binary signal. The experimental set-up is presented in Fig. 3.28. The silicon device (MOSFET) depending on the applied gate-source voltage behaves like a variable resistor (load). Basically, by changing the gate-source voltage we indirectly excite the cell voltage.

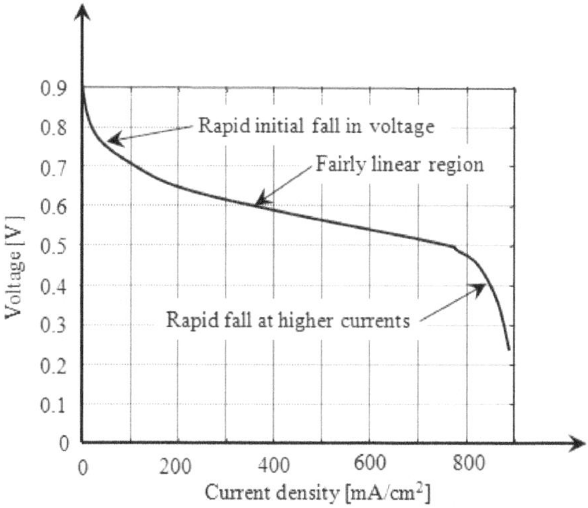

Fig. 3.27 Static characteristic of the fuel cell.

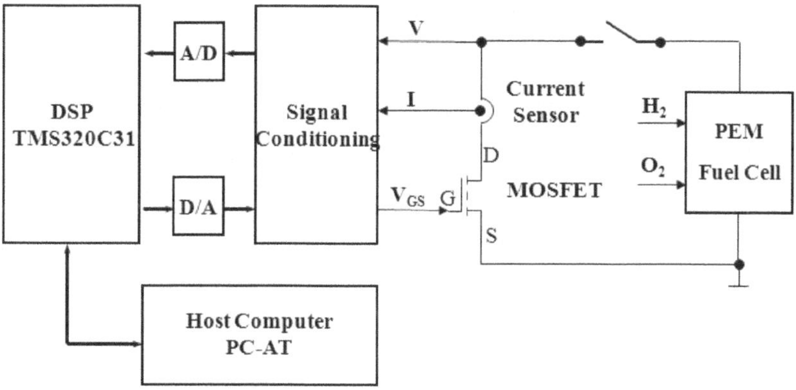

Fig. 3.28 The experimental set-up.

As in the previous example we choose as input signal a 7th-order PRBS as a driving signal for the MOSFET device and the generated current $I(s)$ and voltage $V(s)$ variations will be used for system identification of the investigated fuel cell.

Defining a frequency range for identification as ω_{id} the sampling time T_s is selected according to equation (3.109). The considered sampling time is $T_s = 0.005s$ and the frequency range for identification is: $12[rad/sec] < \omega_{id} < 250[rad/sec]$.

Measurements have been performed for two different operating points (noted by 1 and 2) corresponding for $V_{GS} = 3.8$V and $V_{GS} = 4$V. The internal impedance is derived according to the following equation:

$$Z(s) = \frac{V_1(s) - V_2(s)}{I_2(s) - I_1(s)} \tag{3.116}$$

In order to derive the impedance of the fuel cell - from system identification point of view - the input signal is the current variation and the output signal is the voltage variation. We consider an ARX model structure and we validate the model as been presented in this chapter.

How well the model (dotted line) fits the measured data (solid line) is shown in Fig. 3.29.

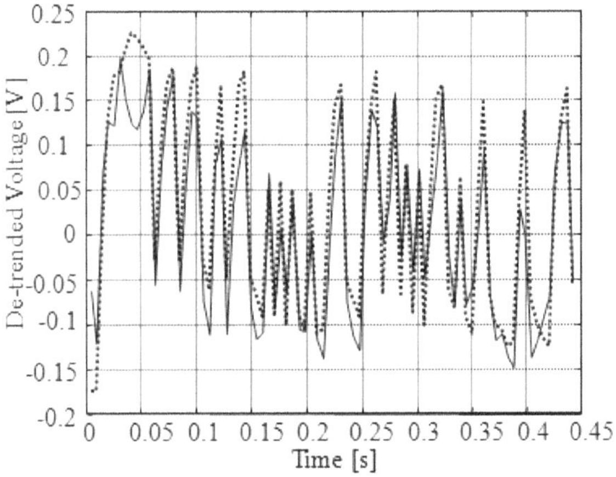

Fig. 3.29 How well the model fits the data?

The derived internal impedance model is given by the next equation:

$$Z_b(s) = K_b \frac{(s - z_1)(s - z_{2,3})(s - z_{4,5})}{(s - p_1)(s - p_{2,3})(s - p_{4,5})} \tag{3.117}$$

where:

$$\begin{aligned} K_b &= 0.8994\Omega \\ z_1 &= -2.178 \cdot 10^2 \\ z_{2,3} &= (-1.68 \pm 2.99i) \cdot 10^2 \\ z_{4,5} &= (-2.36 \pm 6.17i) \cdot 10^2 \\ p_1 &= -1.266 \cdot 10^2 \\ p_{2,3} &= (-1.63 \pm 3.23i) \cdot 10^2 \\ p_{4,5} &= (-2.42 \pm 6.28i) \cdot 10^2 \end{aligned} \tag{3.118}$$

The previously obtained model is a relatively high-order model, therefore model-order reduction is performed using balanced truncation techniques (based on controllability and observability gramiams) as well as based on the physical model [31]. The approach presented here can be seen as a gray box identification approach.

The reduced-order model is given by equation (3.119) and a comparison between the high-order and reduced-order model is presented in Fig. 3.30, where the dashed lined corresponds to the high-order model and the solid line corresponds to the reduced-order model.

$$Z_r(s) = K_r \frac{s - z}{s - p} \tag{3.119}$$

where:

$$\begin{aligned} K_r &= 0.9077\Omega \\ z &= 160.90 \\ p &= -108.62 \end{aligned} \tag{3.120}$$

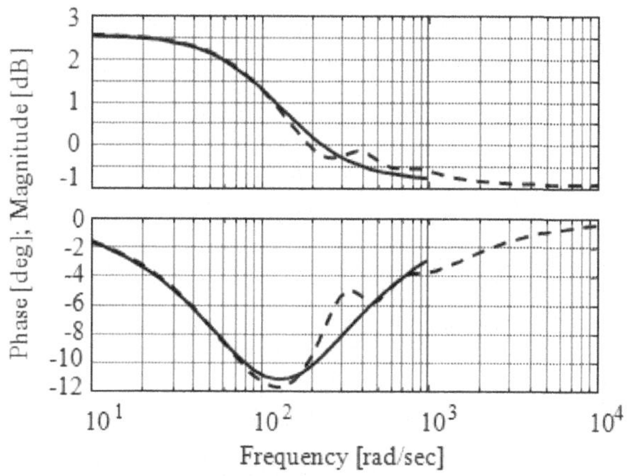

Fig. 3.30 Comparison between high and low-order models.

The last step is to derive the values of the physical parameters of the fuel cell's equivalent circuit (gray box identification). Based on the equivalence between the reduced-order model and equivalent circuit model is possible to derive the parameters of the fuel cell. The obtained parameter values, according to the physical model are:

$$\begin{aligned} R_r &= 0.9077\Omega \\ R_a &= 0.43690\Omega \\ C_a &= 0.0211F \end{aligned} \tag{3.121}$$

The last section is dedicated to a brief review and some concluding remarks related to system identification.

3.6.3 Short Review and Remarks

At the end of this chapter we briefly review the main steps of the system identification, which are summarized in Table 3.3.

Table 3.3 The main steps of system identification for control.

Step 1	Design of experiment	Input signal design, sampling frequency, etc.
Step 2	Experiment execution	Collecting I/O signals
Step 3	Data pre-processing	Eliminating biases, trends, outliers, etc.
Step 4	Structural identification	Selection of model structure, model-order, etc.
Step 5	Parameter estimation	Parameters are estimated using e.g. LSM.
Step 6	Model validation and model-order reduction	Comparison of measured and estimated output, pole-zero cancellation, model-order reduction, etc.
Step 7	Model validation for control system design	Deriving nominal plant and plant uncertainties.

Remarks about System Identification

Pre-processing Data

After the data is collected it is likely that is not ready for immediate use in identification algorithms. There are several possible deficiencies in the data such as:

- high-frequency disturbances in the data record, above the frequency of interest
- drift and offset, low-frequency disturbances with periodic character
- occasional burst and outliers, missing data, non-continuous data records.

Pre-processing the data is an important pre-requisite for the estimation phase. This is accomplished primarily by low-pass or high-pass pre-filtering and/or subtracting offsets and trends from the data. Notice that if the data are pre-filtered the noise model is affected [51].

Drifts and De-trending

Low frequency disturbances, offsets, trends, drift and periodic (seasonal) variations are common in data. There are basically two different approaches to deal with such problems:

- removing the disturbances by explicit pre-treatment of the data (see `dtrend` algorithm in Matlab)
- letting the noise model take care of disturbances.

Outliers and Missing Data

In the recorded data it may be that, simple values or portions are missing or contain non-relevant information. In these cases it is natural to select segments of the original data set, which are considered to contain relevant information about dynamics of interest.

It is also very common situation in practice that a number of separate experiments have been performed. Merging the data sets it's one approach, but we cannot concatenate the data segments, because the connection points would cause transients that may destroy the estimate.

Robust Estimation

The least-squares estimate is optimal if the disturbances are Gaussian and such that the estimation error is white noise. In practice the least-squares estimate has some drawbacks because the assumptions are violated. For examples - outliers - a single large error will have a drastic influence on the result because the errors are squared in the criterion.

One approach for robust estimation is to consider an error function such as:

$$f(\epsilon) = \frac{\epsilon}{1 + a|\epsilon|} \tag{3.122}$$

where a is a positive constant.

The function $f(\epsilon)$ is linear for small errors, but increases more slowly than the linear function for large errors.

Identification in Closed-Loop: Identifiability

It is sometimes necessary to perform the identification experiment under output feedback, i.e. in closed-loop. The reason may be that the plant is unstable, or it has to be controlled for production, economic or safety reasons or that it contains inherent feedback mechanism [51].

The basic problem with closed-loop data is that, typically has less information about the open-loop system - an important purpose of feedback is to make the closed-loop system less sensitive to changes than the open-loop system. Several methods that give consistent estimates for open-loop data may fail when applied in a direct way to closed-loop identification.

Some hints related to closed-loop identification:

- the closed-loop experiments may be non-informative even if the input is persistently exciting. The reason then is that the regulator is too simple

- for open-loop data, output error models will give consistent estimates of the plant, even if the additive noise is not white noise. This is not true for closed-loop data
- the subspace methods will typically not give consistent estimates when applied to closed-loop data.

An interesting result is that, even if no extra input is allowed, informative experiments result if we shift between different linear regulators. This subject is also challenging, but is not discussed here - the interested reader should see [51].

Finally, we close this chapter with a few remarks.

Remarks

Control design is one of the most important uses of the identified models. Frequently, 80% of control design time is spent for model building. It is important to build our model from observed (measured) data and validate the model in time and frequency domain.

The input signal is persistently exciting in a limited frequency range. Therefore, the model can be validated only in that frequency range. If the designed input signal does not cover our frequency range of interest, the identification experiment shall be repeated in different frequency ranges and the models shall be combined in one model.

If the derived model is used in controller design, the confidence "faith", which we have in the designed controller depends on the confidence, which we have in our model. It is true that that we can have good control performance even with a mediocre model, as long as it is reliable in certain frequency ranges. However, this is mainly due to the main characteristics of the feedback control (described in the previous chapter).

The model shall "catch" the main aspects (characteristics) of the plant, which is going to be controlled. If certain aspects are missing from the model it is very likely the control system will not fulfill the requirements. For example, if parameters vary in large limits with temperature and these temperature dependences are not included into the model then the system performance will depend on the operating temperature.

In case of sensorless control the output variable or state variable is estimated based on the other measured variables (outputs) making use of the model. In such a case an accurate model is essential. Sensorless control is a very challenging topic in the industry due to its benefits such as: elimination of sensor(s), cost reduction, easy installation, etc.

In this chapter we have seen how to derive (from measured data) a mathematical model for control, which comprises a low-order model and a plant uncertainty. Therefore, in the next chapters we will focus on model based controller design.

Chapter 4

Classical Controller Design - Part I

'You do not really understand something unless you can explain it to your grandmother.' — Albert Einstein

Summary:

This chapter is dedicated to controller design in a classical framework. The following items are discussed:

- controller design based on pole-zero cancellation and the influence of controller zero
- controller design for deadbeat response
- controller design using the root locus technique
- PID controller tuning
- controller design using coefficient diagram method
- control system validation using Monte Carlo simulation.

In this chapter very simple control design techniques are presented. Most of the methods presented here can be followed using paper and pencil. However, simple software tools for simulation of dynamical systems are recommended, in order to verify the design. This chapter has been developed keeping in mind that not all readers have access to advanced software tools used in controller design.

4.1 Controller Design Based on Pole-Zero Cancellation

In this section a very simple control design procedure is presented, which can be applied for stable plants with poles relatively far from the imaginary axis. Basically, we will cancel the plant's poles by the controller's zeros. Let us illustrate this approach by an example.

Example 4.1. *Let us consider that the plant transfer function is given by:*

$$P(s) = \frac{1}{(s+1)(s+2)} \tag{4.1}$$

and the performance specifications are formulated in terms of settling time $T_{set} = 1.2s$ *and percent of overshoot* $P.O. \leq 4\%$.

First, the time-domain specifications are translated into frequency domain, which leads to $\xi = 0.707$ and $\omega_n = 4.71$.

According to the pole-zero cancellation technique, we are going to cancel the plant's poles with the controller's zeros. Therefore, we are looking for a controller having the form:

$$K(s) = k\frac{(s+z_1)(s+z_2)}{s(s+p)} \tag{4.2}$$

where $z_1 = 1$ and $z_2 = 2$.

The closed-loop transfer function (considering unity feedback) can be written as:

$$T(s) = \frac{P(s)K(s)}{1+P(s)K(s)} = \frac{k}{s^2+ps+k} \tag{4.3}$$

The characteristic equation of the closed-loop system is made identical with the desired characteristic equation:

$$q(s) = s^2 + ps + k \equiv s^2 + 2\xi\omega_n s + \omega_n^2 \tag{4.4}$$

then using coefficient identification techniques we obtain $k = 22.2$ and $p = 6.66$.

The controller is given by:

$$K(s) = 22.2\frac{(s+1)(s+2)}{s(s+6.66)} \tag{4.5}$$

The system's step response is presented in Fig. 4.1 and the compensator leads to gain margin $G_m = 2.77 \cdot 10^5$ and phase margin $\phi_m = 65°$, shown in Fig. 4.2.

4.1.1 The Influence of Controller Zero

A frequently used compensator has the form:

$$K(s) = k\frac{s+z}{s+p} \tag{4.6}$$

that alter the roots of the characteristic equation of the closed-loop system. However, the closed-loop transfer function, $T(s)$ will contain the zero of $K(s)$ as zero of $T(s)$. This zero might significantly affect the response of the system $T(s)$.

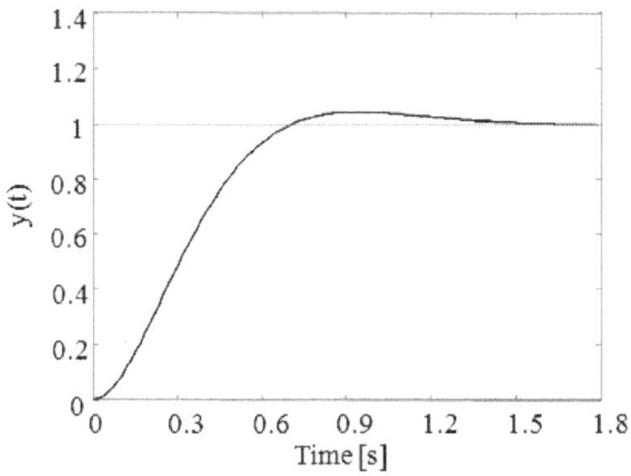

Fig. 4.1 Step response.

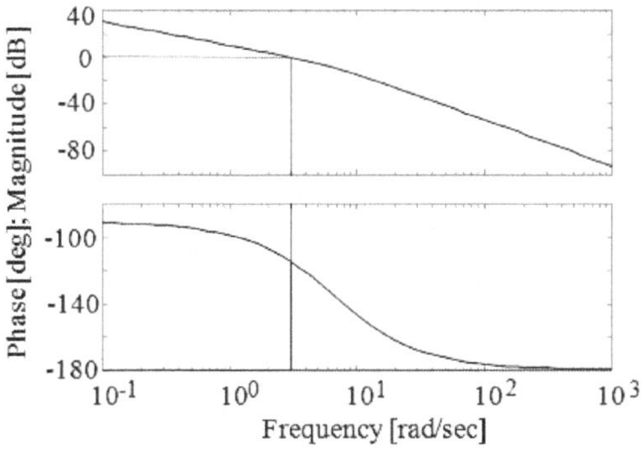

Fig. 4.2 Gain and phase margins.

Example 4.2. *Let us consider the system:*

$$P(s) = \frac{1}{s} \tag{4.7}$$

We will introduce a PI compensator so that:

$$K(s) = k_1 + \frac{k_2}{s} = \frac{k_1 s + k_2}{s} \tag{4.8}$$

The closed-loop transfer function of the system is:

$$T(s) = \frac{k_1 s + k_2}{s^2 + k_1 s + k_2} \tag{4.9}$$

If we require $T_{set} = 0.5$ and P.O. $= 4\%$ then:

$$\xi = 0.707 \ \xi\omega_n = 8 \ \omega_n = 8\sqrt{2} \tag{4.10}$$

Thus, using coefficient identification techniques we obtain:

$$\begin{aligned} q(s) = s^2 + k_1 s + k_2 &= s^2 + 2\xi\omega_n s + \omega_n^2 \\ k_1 = 2\xi\omega_n &= 16 \\ k_2 = \omega_n^2 &= 128 \end{aligned} \tag{4.11}$$

where $q(s)$ denotes the characteristic equation.

The closed-loop transfer function is:

$$T(s) = \frac{16(s+8)}{s^2 + 16s + 128} \tag{4.12}$$

The effect of the zero on the step response is significant (the overshoot of the step response is 21%). Therefore, we use a prefilter, also called set-point filter or command smoother (shown in Fig. 4.3) $K_p(s)$ to eliminate the zero from $T(s)$, while maintaining the dc gain 1, thus:

$$K_p(s) = \frac{8}{s+8} \tag{4.13}$$

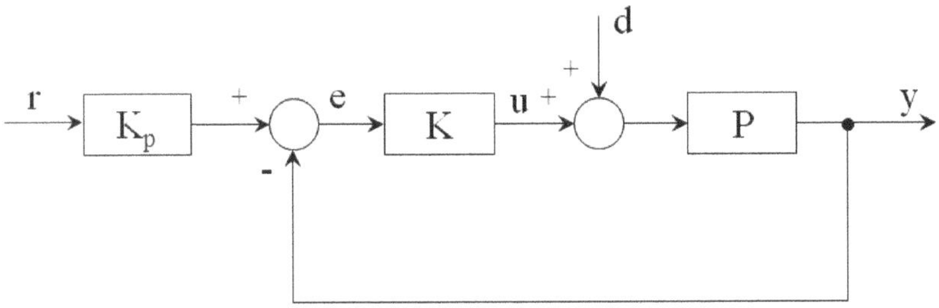

Fig. 4.3 Control block diagram with prefilter.

Then we have:

$$T(s) = \frac{128}{s^2 + 16s + 128} \tag{4.14}$$

and the overshoot of this system is 4.5%, as expected, shown in Fig. 4.4. It is possible the prove that the zero at $s = -a$ has a significant effect when $a/\xi\omega_n < 5$, where -a is the zero, and $-\xi\omega_n$ is the real part of the dominant roots of the characteristic equation of $T(s)$.

We remark, that pole-zero cancellation is not a robust scheme, since it exactly cancels the plant's poles with the controller's zeros. Since the plant poles can vary - due to parameter variations - it is recommended to place the controller's zeros a little bit left to the poles, which are going to be canceled. However, approximate pole-zero cancellation will give us many of the desirable characteristics.

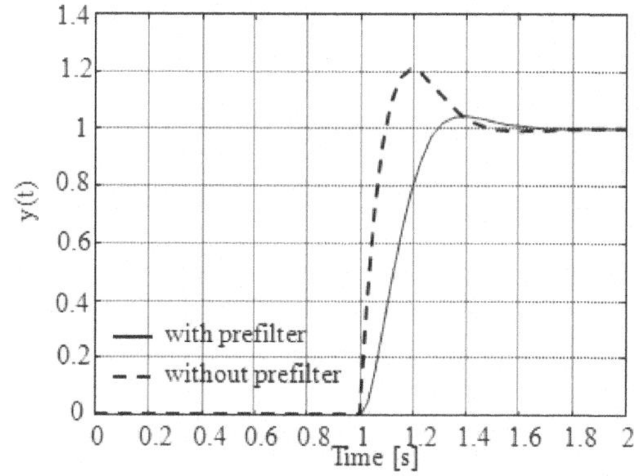

Fig. 4.4 The influence of prefilter.

4.2 Controller Design for Deadbeat Response

Often the goal for a control system is to achieve a fast response to a step command with minimal overshoot. We define the deadbeat response as a response that proceeds rapidly to the desired level and holds at that level with minimal overshoot.

We use the ±2% band at the desired level as acceptable range of variation from the desired response. If the system response enters the band at time T_{set}, it has satisfied the settling time T_{set} upon entry to the band.

A deadbeat response has the following characteristics [23]:

- steady-state error = 0;
- fast response - minimum rise time and settling time;
- *P.O.* < 2%;
- percent of undershoot < 2%.

The parameters of the closed-loop transfer function which assure a deadbeat response (for different system's order) are taken from [23] and are summarized in Table 4.1.

Table 4.1 Coefficients and response measures of a deadbeat system.

System order	α	β	γ	δ	ε	P.O. [%]	T_{setn}
2nd	1.82					0.10	4.82
3rd	1.90	2.20				1.65	4.04
4th	2.20	3.50	2.80			0.89	4.81
5th	2.70	4.90	5.40	3.40		1.29	5.43
6th	3.15	6.5	8.70	7.55	4.05	1.63	6.04

Therefore, the characteristic equation of a 6th-order deadbeat system will be:

$$q(s) = s^6 + \alpha\omega_n s^5 + \beta\omega_n^2 s^4 + \gamma\omega_n^3 s^3 + \delta\omega_n^4 s^2 + \varepsilon\omega_n^5 s + \omega_n^6 \tag{4.15}$$

Note: In Table 4.1 the settling time T_{set} is normalized to $T_{setn} = \omega_n T_{set}$.

Example 4.3. *Let us consider a unity feedback system with compensator* $K(s)$ *and prefilter* $K_p(s)$*. The plant is:*

$$P(s) = \frac{1}{(s+1)(s+2)} \tag{4.16}$$

In this case we are looking for a compensator having the form:

$$K(s) = k\frac{(s+z_1)(s+z_2)}{s(s+p)} \tag{4.17}$$

which will lead to a 4th-order characteristic equation with four unknown parameters.

Moreover, we will consider a prefilter having the form:

$$K_p(s) = \frac{z_1 z_2}{(s+z_1)(s+z_2)} \tag{4.18}$$

The transfer function of a closed-loop system is noted with $T(s)$, in order to determine the coefficients that yield the optimal deadbeat response, the standard transfer function (4.19) is first normalized:

$$T(s) = \frac{\omega_n^4}{s^4 + \alpha\omega_n s^3 + \beta\omega_n^2 s^2 + \gamma\omega_n^3 s + \omega_n^4} \tag{4.19}$$

Let note with $\bar{s} = s/\omega_n$

$$T(s) = \frac{1}{\bar{s}^4 + \alpha\bar{s}^3 + \beta\bar{s}^2 + \gamma\bar{s} + 1} \tag{4.20}$$

The closed-loop transfer function becomes:

$$T(s) = \frac{kz_1z_2}{s(s+p)(s+1)(s+2) + k(s+z_1)(s+z_2)} \tag{4.21}$$

$$T(s) = \frac{kz_1z_2}{s^4 + (3+p)s^3 + (3p+k+2)s^2 + (2p+kz_1kz_2)s + kz_1z_2} \tag{4.22}$$

For a fourth-order system, if we require settling time $T_{set} = 1.2s$, then $T_{setn} = \omega_n T_{set} = 4.81$, which leads to $\omega_n = 4$.

The characteristic equation of the closed-loop system is:

$$q(s) = s^4 + \alpha\omega_n s^3 + \beta\omega_n^2 s^2 + \gamma\omega_n^3 s + \omega_n^4 = s^4 + 8.8s^3 + 56s^2 + 179.2s + 256 \tag{4.23}$$

then using coefficient identification techniques the following system is obtained:

$$\begin{cases} 3+p=8.8 \\ 3p+2+k=56 \\ 2p+kz_1+kz_2=179.2 \\ kz_1z_2=256 \end{cases} \tag{4.24}$$

Solving the previous system, we obtain the following compensator:

$$K(s)=k\frac{(s+z_1)(s+z_2)}{s(s+p)}=36.6\frac{s^2+4.58s+7}{s(s+5.8)} \tag{4.25}$$

with the prefilter:

$$K_p(s)=\frac{z_1z_2}{(s+z_1)(s+z_2)}=\frac{7}{s^2+4.58s+7} \tag{4.26}$$

The system's step response is presented in Fig. 4.5, the deadbeat compensator leads to gain margin $G_m=1.68\cdot10^5$ and phase margin $\phi_m=30°$, shown in Fig. 4.6.

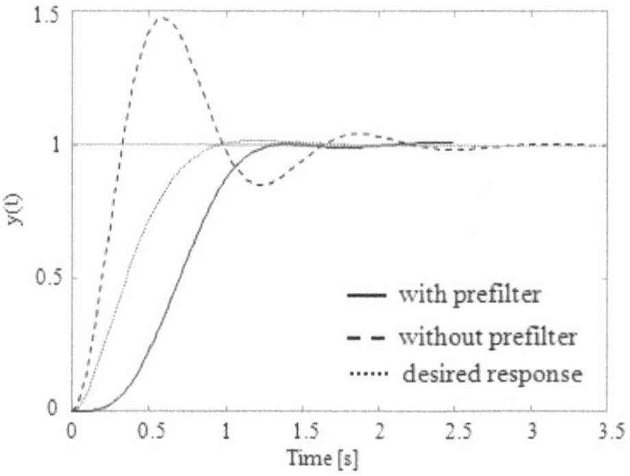

Fig. 4.5 Step response.

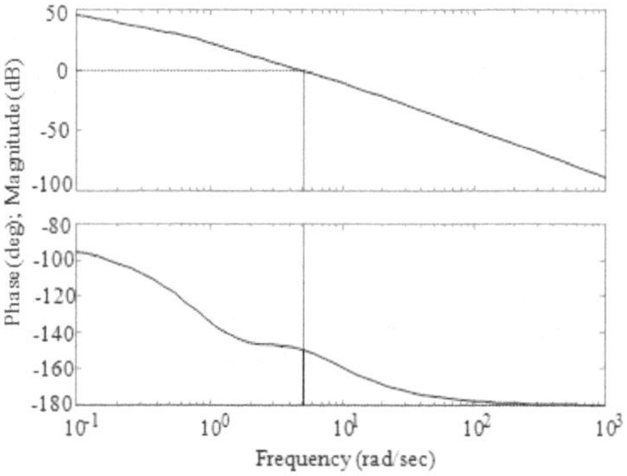

Fig. 4.6 Gain and phase margins.

4.3 Controller Design Using the Root Locus Technique

This section deals with controller design using the root locus technique [23]. The relative stability and the transient performance of a closed-loop control system are directly related to the location of the closed-loop roots of the characteristic equation in the s-plane. It is frequently necessary to adjust one or more system parameters in order to obtain suitable root locations. Therefore, it is useful to determine the locus of the roots in the s-plane as a parameter is varied.

First, let us consider a common example where the plant and controller are expressed in pole-zero forms as:

$$P(s) = k_p \frac{(s - z_{i1})...(s - z_{im})}{(s - p_{i1})...(s - p_{in})} \tag{4.27}$$

$$K(s) = k_k \frac{(s - z_{j1})...(s - z_{jm})}{(s - p_{j1})...(s - p_{jn})} \tag{4.28}$$

with $k = k_p k_k$

$$\begin{aligned} N(s) &= (s - z_{i1})...(s - z_{im})(s - z_{j1})...(s - z_{jm}) \\ D(s) &= (s - p_{i1})...(s - p_{in})(s - p_{j1})...(s - p_{jn}) \end{aligned} \tag{4.29}$$

To simplify the notations set $m = i_m + j_m$ and $n = i_n + j_n$, then $G(s) = K(s)P(s)$ is written as:

$$G(s) = k \frac{(s - z_1)...(s - z_m)}{(s - p_1)...(s - p_n)} = k \frac{N(s)}{D(s)} \tag{4.30}$$

Assuming $k \neq 0$ the characteristic equation is equivalent to:

$$1 + G(s) = 0 \equiv 1 + kG_1(s) = 0 \tag{4.31}$$

where $G_1(s) = N(s)/D(s)$

The purpose of the root locus is to examine how the closed-loop system poles changes as k varies from 0 to ∞. Clearly, a point $r \in \mathcal{C}$ is on the root locus iff (if and only if):

$$k = -\frac{1}{G_1(r)} \tag{4.32}$$

This condition can be separated into two parts:

$$|k| = \frac{1}{|G_1(r)|} \tag{4.33}$$

$$\angle k = 0^\circ = -(2l + 1)180^\circ - \angle G_1(r) \; l = 0, \pm 1, \pm 2, ... \tag{4.34}$$

The phase rule determines the points in the complex plane $\mathcal{C}$ that are on the root locus. The magnitude determines the gain $k > 0$ for which the root locus is at a given point r. By using the definition of $G_1(s)$, the phase condition can be rewritten as:

$$(2l+1)180^\circ = \sum_{i=1}^{n} \angle(r-p_i) - \sum_{j=1}^{m} \angle(r-z_j) \tag{4.35}$$

Similarly, the magnitude rule is equivalent to:

$$k = \frac{\Pi_{i=1}^{n}|r-p_i|}{\Pi_{j=1}^{m}|r-z_j|} \tag{4.36}$$

The phase rule is rewritten:

$$-\sum_i \theta_{p_i} + \sum_j \theta_{z_j} = -(2l+1)180^\circ \tag{4.37}$$

the angles for poles θ_p and for zeros θ_z, are all measured in counter-clockwise direction from the horizontal line to the desired root location.

By adding a compensator, we would like to move the dominant poles of the closed-loop system (roots of the characteristic equation) to the desired root locations.

Time domain performance specifications are translated into terms such as: damping ξ and natural frequency ω_n as we have seen in chapter 2. Then the desired root locations are defined as the roots of a second-order system:

$$r_{1,2} = -\xi\omega_n \pm \omega_n\sqrt{(\xi^2-1)} \tag{4.38}$$

Therefore, the poles of $T(s)$ should lie in a region of the complex plane defined by the minimum allowable ξ and $\xi\omega_n$, the dashed area shown in Fig. 4.7.

If the order of the closed-loop transfer function is higher than two, depending on the other poles and zeros it might be possible to approximate the system with a second-order system.

The third-order closed-loop transfer function $T(s)$ behaves like a second-order system. The effect of the third root $r_3 = -\alpha$ is negligible if $\alpha \gg \xi\omega_n$.

$$T(s) = \frac{\alpha\omega_n^2}{(s^2+2\xi\omega_n s+\omega_n^2)(s+\alpha)} \tag{4.39}$$

Next, we are going to alter the root locations by choosing a first-order compensator $K(s)$ with the transfer function:

$$K(s) = k\frac{s+z}{s+p} \tag{4.40}$$

- when $|z| < |p|$, the network is called a phase-lead network, shown in Fig. 4.8 a.
- if $|p| < |z|$ the network is called a phase-lag network, shown in Fig. 4.8 b.

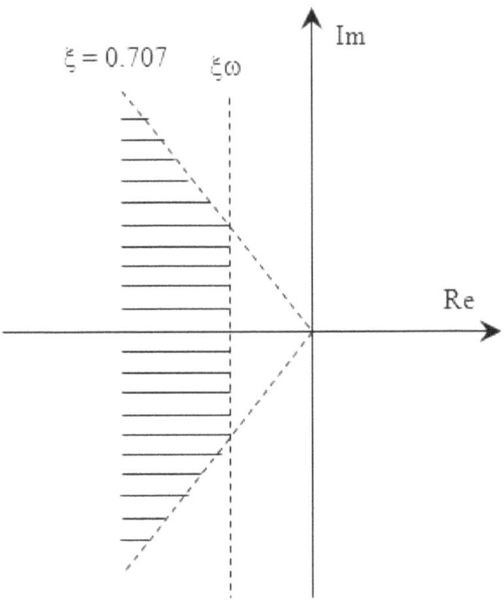

Fig. 4.7 Region of the desired closed-loop poles.

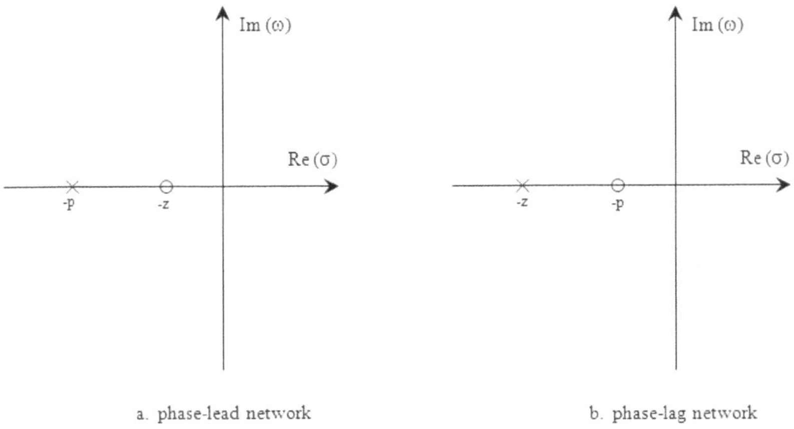

Fig. 4.8 Lead-lag networks.

We remark that, there are two independent conditions (magnitude and phase), which can be used to find out the controller parameters. If the controller has let's say three parameters (i.e. $K(s)$), during design we have to fix one of them and find out the other two, based on the magnitude and phase condition. For example, after placing at the desired location the controller zero or pole based on phase condition the overall loop gain can be set to satisfy the steady-state error requirement.

4.3.1 Phase-Lead Controller Design Using the Root Locus

This section deals with phase-lead network design. The main design steps are summarized as follows and are illustrated by examples [23].

- translate the system specifications into desired root location for the dominant roots
- sketch the uncompensated root locus, find out if the desired root locations can be realized with an uncompensated system
- if a compensator is necessary, place the zero of the phase-lead network directly below the desired root location (or to the left of the first two real poles)
- find out the pole location so that the total angle at the desired root location is -180°
- evaluate the total system gain at the desired root location and then calculate the error constant
- repeat the steps if the error constant is not satisfactory.

Example 4.4. *Let us consider that the plant transfer function is given by equation (4.41) and the performance specifications are formulated in terms of settling time $T_{set} = 0.5s$ and percent of overshoot P.O. $\leq 10\%$.*

$$P(s) = \frac{1}{(s+1)(s+2)} \tag{4.41}$$

we are looking for a compensator having the form $K(s) = k(s+z)/(s+p)$.

The performance specifications given in time domain are transformed into performance specifications in frequency domain, which leads to $\xi = 0.59$ and $\omega_n = 13.49$, thus the desired root locations are:

$$r_{1,2} = -\xi\omega_n \pm \omega_n\sqrt{\xi^2 - 1} = -8 \pm 10.8j \tag{4.42}$$

Next, the simplest design procedure is to place the zero of the phase lead network directly below the desired root location (shown in Fig. 4.9), that means $z = 8$. Furthermore, we find out the pole location so that the total angle at the desired root location is -180°.

We calculate (using elementary geometry) the angles θ_{p1} and θ_{p2}.

$$\begin{aligned} tg(180° - \theta_{p1}) &= tg(10.8/6) = 1.8 = tg(61°) \\ \theta_{p1} &= 180° - 61° = 119° \end{aligned} \tag{4.43}$$

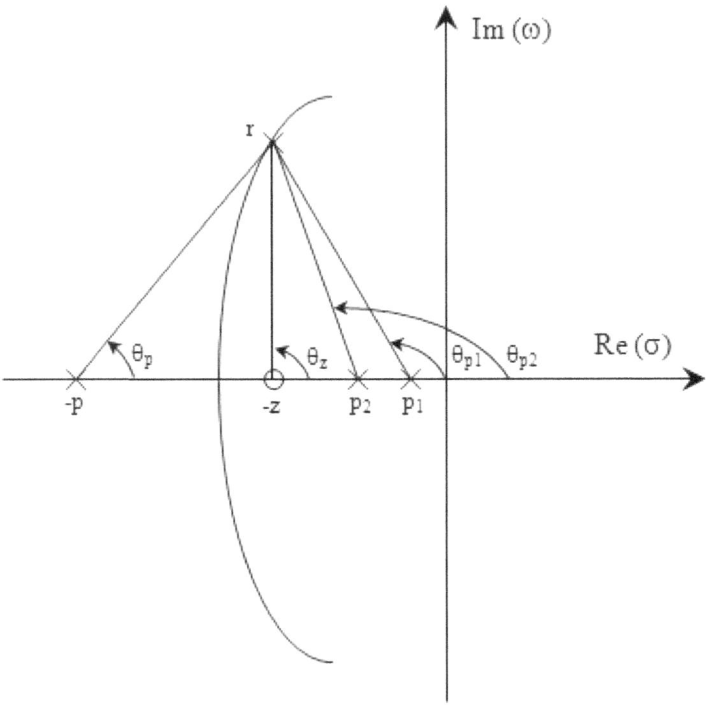

Fig. 4.9 Root locus - phase-lead network design.

$$\begin{aligned} &tg(180° - \theta_{p2}) = tg(10.8/7) = 1.54 = tg(57°) \\ &\theta_{p2} = 180° - 57° = 123° \end{aligned} \tag{4.44}$$

and then the total angle at the desired root location $theta_p$.

$$\begin{aligned} \theta_p &= 180° + \theta_z - \theta_{p1} - \theta_{p2} \\ \theta_p &= 180° + 90° - 119° - 123° = 29° \end{aligned} \tag{4.45}$$

Moreover, using elementary geometry we compute the location of the compensator pole:

$$p = -Re(r_{1,2}) + \frac{|Im(r_{1,2})|}{\tan \theta_p} = 28.16 \tag{4.46}$$

Then, we evaluate the total system gain at the desired root location and then calculate the error constant. The obtained gain value is $k = 338$ according to equation (4.36):

$$k = \frac{|-8 + 10.8j - (-1)||-8 + 10.8j - (-2)||-8 + 10.8j - (-28.16)|}{|-8 + 10.8j - (-8)|} = 338 \tag{4.47}$$

The compensator is:

$$K(s) = k\frac{s+z}{s+p} = 338\frac{s+8}{s+28.16} \tag{4.48}$$

which leads to gain margin $G_m = 1.8 \cdot 10^4$ and phase margin $\phi_m = 46°$.

The next step is to compute the steady state error in the case of unit step. According to the final value theorem we have:

$$e_{ss} = \lim_{s\to 0} sE(s) = \lim_{s\to 0} s\frac{1}{1+P(s)K(s)}R(s) = \lim_{s\to 0} s\frac{1}{1+P(s)K(s)}\frac{1}{s} \approx 0.02 \tag{4.49}$$

Finally, in order to eliminate the effect of the compensator zero we utilize a prefilter:

$$K_p(s) = \frac{z}{s+z} = \frac{8}{s+8} \tag{4.50}$$

The system's root locus is shown in Fig. 4.10 and the closed-loop system transient response is presented in Fig. 4.11.

The advantage of the root locus method is the ability of the designer to specify the location of the dominant roots and therefore the transient response. The disadvantage of the method is that one cannot directly specify an error constant.

Furthermore, caution is necessary when the zero is placed: the zero must not alter the dominance of the desired roots; that is the zero should not be placed nearer the origin than the second pole on the real axis (the real root near the origin will dominate the system response).

As conclusion, the specifications for the transient response of this system were originally expressed in terms of the overshoot and the settling time of the system. These specifications were translated, on the basis of approximating

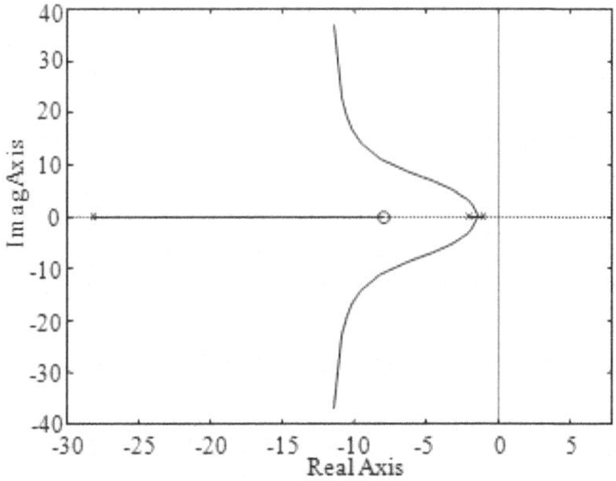

Fig. 4.10 Root locus.

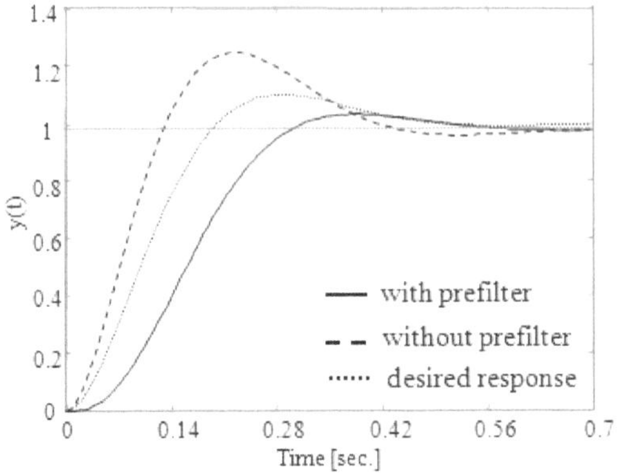

Fig. 4.11 Step response.

the system by a second-order system, to an equivalent ξ and ω_n and therefore a desired root location. However, the original specifications will be satisfied only if the roots selected are dominant. Moreover, to eliminate the effect of the zero in the closed-loop transfer function we utilized a prefilter.

4.3.2 Phase-Lag Controller Design Using the Root Locus

In many cases the primary objective of control is to obtain a high steady-state accuracy, therefore integration-type networks are used.

One form of controller widely used is the proportional-integral (PI) controller

$$K(s) = k_P + \frac{k_I}{s} = k_P \frac{s + k_I/k_P}{s} \tag{4.51}$$

The main steps of design are summarized as follows and a possible root locus is shown in Fig. 4.12.

- translate the system specifications into desired root location for the dominant roots ($r_1 = p_1$ and $r_2 = p_2$)
- sketch the uncompensated root locus, find out if the desired root locations can be realized with an uncompensated system
- if a compensator is necessary, place the pole of the integration network at the origin ($p = 0$).
- find out the zero location (z) so that the total angle at the desired root location is -180°
- evaluate the total system gain at the desired root location.

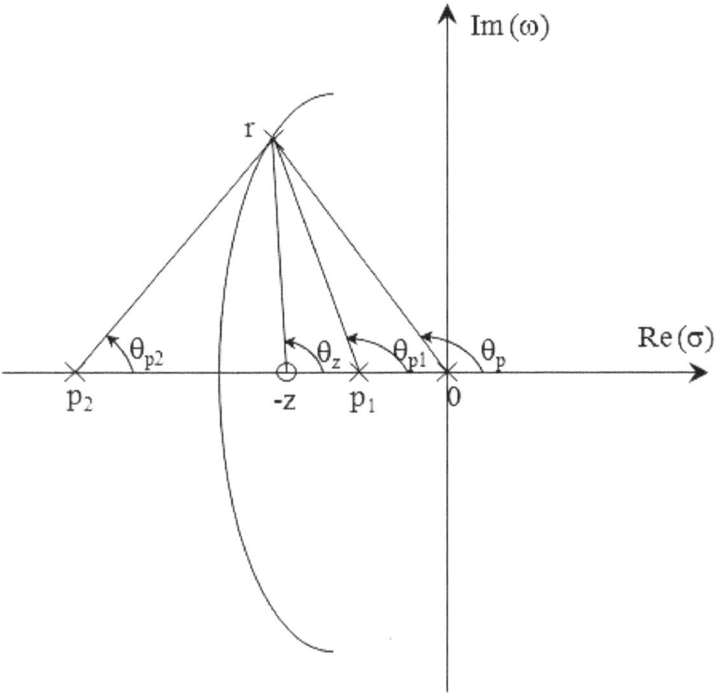

Fig. 4.12 Root locus - integration network design.

Example 4.5. *The transfer function of a temperature control system is:*

$$P(s) = \frac{1}{(2s+1)(0.5s+1)} \tag{4.52}$$

To maintain zero steady-state error for a step input we will use a PI controller:

$$K(s) = K_P + \frac{K_I}{s} = K_P\frac{s+z}{s} \tag{4.53}$$

and we will require P.O. less than 10% and settling time $T_{set} = 5s$.

The desired root locations are at: $r_{1,2} = -0.8 \pm 1.08j$.

After we place the compensator pole at the origin, we will determine the location of the zero, by ensuring that the total angle at the desired root location is -180°.

$$\begin{aligned} &\sum \theta_z = -180^\circ + \sum \theta_p \\ &\theta_z = -180^\circ + 273.8^\circ = 93.8^\circ \end{aligned} \tag{4.54}$$

Moreover, using elementary geometry we compute the location of the compensator zero:

$$z = -Re(r_{1,2}) + \frac{|Im(r_{1,2})|}{\tan\theta_z} = 0.72 \tag{4.55}$$

Finally, to determine the gain at the desired root we evaluate the vector lengths from poles and zeros according to equation (4.36) and obtain $k = 2.262$.

The controller transfer function is:

$$K(s) = k\frac{s+z}{s} = 2.262\frac{s+0.72}{s} \tag{4.56}$$

The effect of the zero is to increase the overshoot to a step input, therefore we use a prefilter so that the zero is eliminated in the closed-loop transfer function by setting:

$$K_p(s) = \frac{0.72}{s+0.72} \tag{4.57}$$

Finally we check the obtained gain and phase margins: $G_m = 2.72 \cdot 10^6$ and $\phi_m = 52.4°$, which is reasonable.

The system's root locus is shown in Fig. 4.13 and the closed-loop system transient response is presented in Fig. 4.14.

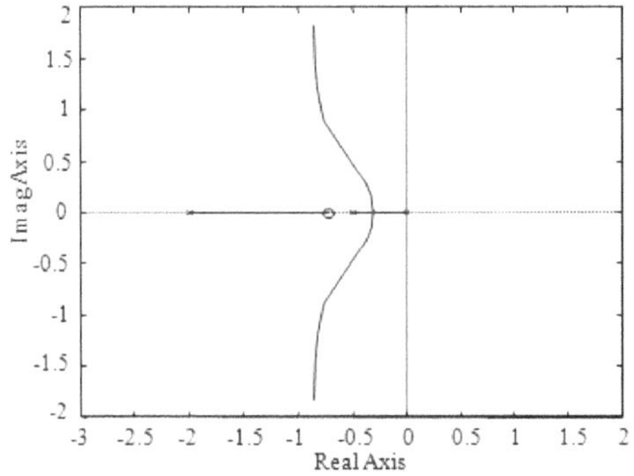

Fig. 4.13 Root locus.

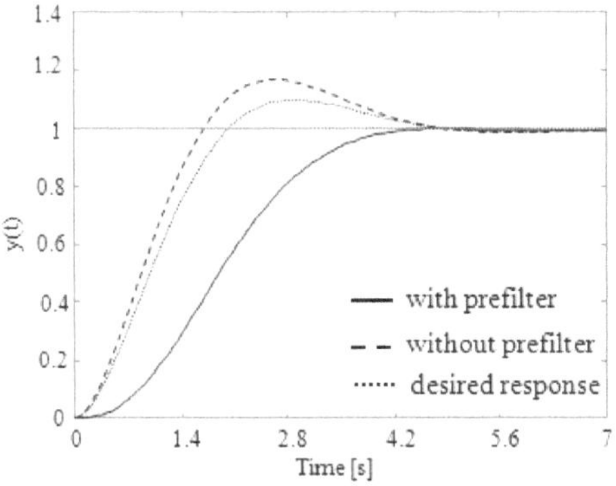

Fig. 4.14 Step response.

The main characteristics of phase-lead and phase-lag compensation networks are summarized in Table 4.2.

Table 4.2 Characteristics of phase-lead and phase-lag compensation networks.

Compensation network	Phase-lead	Phase-lag
Results	Increases system bandwidth Increases gain at higher frequencies	Decreases system bandwidth
Advantages	Improves dynamic response	High-frequency noise is suppressed Steady-state error reduced
Disadvantages	Increases bandwidth and thus susceptibility to noise	Slows down transient response
Applications	When fast transient response is desired	When steady-state error is specified

4.4 PID Controller Design

PID (proportional integral derivative) control due to its simple structure is widely used in the industry. In this section very briefly we recall some well-known tuning techniques. The transfer function of the PID controller is written as:

$$K(s) = K_p \left(1 + \frac{1}{sT_{int}} + sT_{der}\right) \tag{4.58}$$

where the three tuning parameters are: K_p, T_{int} and T_{der}.

The above transfer function is not a proper transfer function, so in practice the PID controller is implemented as:

$$K(s) = K_p \left(1 + \frac{1}{sT_{int}} + \frac{sT_{der}}{1 + \tau s}\right) \tag{4.59}$$

Basically, the derivative term is low-pass filtered.

4.4.1 Ziegler-Nichols Tuning Formula

A very useful empirical tuning formula was proposed in early 1942 by Ziegler and Nichols. The tuning formula is mainly for a first-order plant with time delay. In practice a large variety of plants can be modeled approximately by the equation below:

$$P(s) = \frac{k}{1 + Ts} e^{-sT_d} \tag{4.60}$$

where the term e^{-sT_d} is the transfer function of the pure delay element.

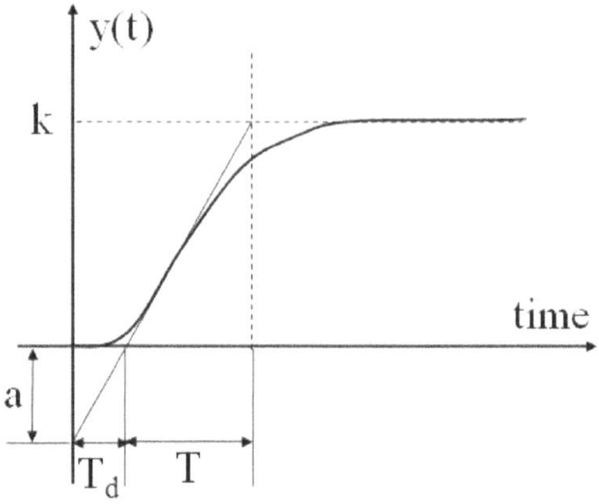

Fig. 4.15 Step response of a first-order system with time delay.

In Fig. 4.15 the step response of a first-order system with time delay is shown.

Based on the measured parameters a, time delay T_d and time constant T the controller parameters are derived. The Ziegler-Nichols tuning formulae are mentioned in the Table 4.3.

Table 4.3 Ziegler-Nichols tuning formulae.

Controller type	K_p	T_{int}	T_{der}
P	1/a		
PI	0.9/a	$3T_d$	
PID	1.2/a	$2T_d$	$T_d/2$

Another approach suggested by Ziegler and Nichols was to control the system using a proportional controller and increase the gain until oscillations appear. Based on the proportional gain and the oscillation frequency is possible to derive the PID controller parameters.

This approach is difficult to apply in practice because in many cases is not safe to increase the gain until oscillations appears or due to saturation of control elements or control signal the system is no longer a linear system. However, if a model is available using numerical simulation the procedure can be applied.

4.4.2 Chien-Hrones-Reswick Tuning Formula

There are many variants of the traditional Ziegler-Nichols PID tuning method. The Chien-Hrones-Reswick formula emphasizes the set-point regulation or disturbance rejection [81]. Compared with the traditional Ziegler-Nichols tuning formula the CHR method uses the time constant T of the plant explicitly. The parameters of the controller - optimized either for set-point regulation or disturbance rejection - are mentioned in the Table 4.4 and Table 4.5.

An example related to the CHR method is given in the next chapter, where two degree-of-freedom (2DOF) controllers are discussed.

Table 4.4 CHR tuning formulae for set-point regulation.

Controller	with 0% overshoot			with 20% overshoot		
type	K_p	T_{int}	T_{der}	K_p	T_{int}	T_{der}
P	0.3/a			0.7/a		
PI	0.35/a	1.2T		0.6/a	T	
PID	0.6/a	T	$0.5T_d$	0.95/a	1.4T	$0.47T_d$

Table 4.5 CHR tuning formulae for disturbance rejection.

Controller	with 0% overshoot			with 20% overshoot		
type	K_p	T_{int}	T_{der}	K_p	T_{int}	T_{der}
P	0.3/a			0.7/a		
PI	0.6/a	$4T_d$		0.7/a	$2.3T_d$	
PID	0.95/a	$2.4T_d$	$0.42T_d$	1.2/a	$2T_d$	$0.42T_d$

4.5 The Coefficient Diagram Method

It is well-known that the designed controller shall fulfill several constraints and requirements such as:

- shall be minimum degree, minimum phase (if possible) and stable
- shall have enough bandwidth
- shall have proper settling time and overshoot
- shall be robust against parameter variations
- shall have good disturbance rejection properties
- the magnitude of control signal shall not exceed the specified limit.

The controller designed using the coefficient diagram method (CDM) fulfills most of the conditions mentioned above.

Although the main principles of CDM have been known since 1950s, the first systematic method was proposed by Shunji Manabe [56]. The standard block diagram of CDM is shown in Fig. 4.16. It is a two-degree-of-freedom (2DOF) control structure having a prefilter (set-point filter).

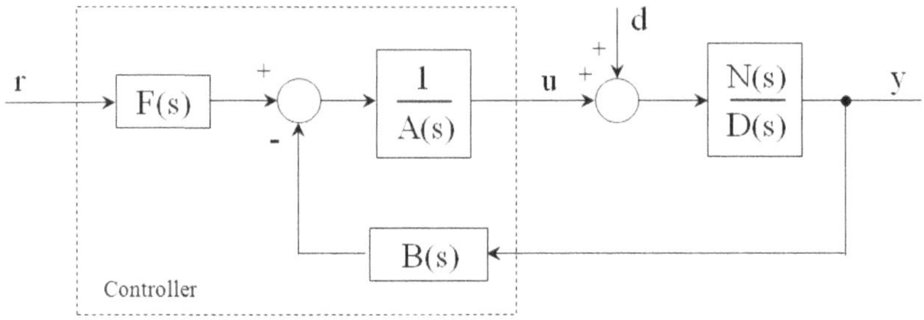

Fig. 4.16 The standard block diagram of CDM.

The controller polynomials $A(s)$ and $B(s)$ appearing in the CDM control structure are:

$$\begin{aligned} A(s) &= \sum_{i=0}^{p} l_i s^i \\ B(s) &= \sum_{i=0}^{q} k_i s^i \end{aligned} \tag{4.61}$$

The output of the CDM is:

$$y = \frac{N(s)F(s)}{q(s)} r + \frac{A(s)N(s)}{q(s)} d \tag{4.62}$$

where the $q(s)$ is the characteristic polynomial of the closed-loop system defined as:

$$q(s) = A(s)D(s) + B(s)N(s) = \sum_{i=0}^{n} a_i s^i \tag{4.63}$$

where a_i are the coefficients of the $q(s)$.

It is easy to show that the block diagram shown in Fig. 4.16 is equivalent with the block diagram shown in Fig. 4.17.

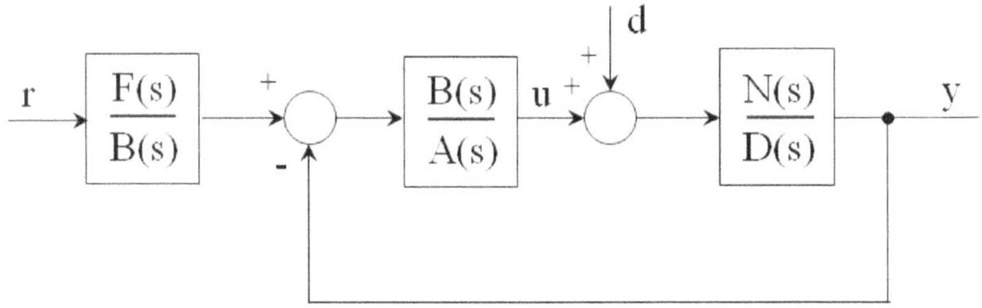

Fig. 4.17 Equivalent block diagram of CDM.

The design parameters of CDM are the equivalent time constant τ, stability index γ_i and stability limits γ_i^*.

The relations between these parameters and the coefficients of the characteristic polynomial are shown below [57, 40]:

$$\tau = a_1/a_0 \tag{4.64}$$

$$\gamma_i = \frac{a_i^2}{a_{i+1}a_{i-1}} \; i = 1, ..., n \; \gamma_0 = \gamma_n = \infty \tag{4.65}$$

$$\gamma_i^* = \frac{1}{\gamma_{i-1}} + \frac{1}{\gamma_{i+1}} \tag{4.66}$$

From the previous equations the coefficients can be written:

$$a_i = \frac{\tau^i}{\Pi_{j=1}^{i-1} \gamma_{i-j}^j} a_0 = Z_i a_0 \tag{4.67}$$

Finally, the target characteristic polynomial is obtained as

$$q(s) = a_0 \left[\left\{ \sum_{i=2}^{n} \left(\Pi_{j=1}^{i-1} \frac{1}{\gamma_{i-j}^j} \right) (\tau s)^i \right\} + \tau s + 1 \right] = a_0 \sum_{i=0}^{n} Z_i s^i \tag{4.68}$$

The instability/stability conditions for the characteristic polynomial of CDM are:

Necessary condition for instability:

$$\gamma_i \gamma_{i-1} \leq 1 \; for \; all \; i, \; i = 2, ..., (n-1) \tag{4.69}$$

Sufficient condition for stability:

$$\gamma_i > 1.1236 \gamma_i^* \; for \; some \; i, \; i = 2, ..., (n-1) \tag{4.70}$$

It is recommended that the standard Manabe form based on Lipatov and Sokolov's conditions is used for CDM design [57, 40].

The stability indices are chosen as:

$$\gamma_i = \{2.5, 2, ..., 2\} \; for \; i = 1, ..., (n-1) \tag{4.71}$$

The most important properties of this form are:

- the step response of the closed loop system usually has no overshoot
- the settling time in case of step input is approximately 2.5τ
- robustness of the control system is acceptable in most cases, when the stability indices are chosen in standard form
- gain and phase margins are obtained in the desired values.

Moreover, designers have freedom in controller design, since they can deviate from the standard form as far as $\gamma_i > 1.5\gamma_i^*$. Another important property is that the settling time T_{set} is defined at the beginning of design. Considering the standard Manabe form, the equivalent time constant is chosen to be $\tau = T_{set}/2.5$. Furthermore, there is an implicit relation between the equivalent time constant and the magnitude of the control signal. Smaller time constants lead to slower time response and implicitly smaller control signal. In practical applications the magnitude of the control signal is always bounded and this shall be taken into account during design.

Next we will illustrate the design procedure by an example.

Example 4.6. *Let us consider the following plant, with gain $K = 0.5$ and time constants $T1 = 0.5$ and $T_2 = 1.5$:*

$$P(s) = \frac{N(s)}{D(s)} = \frac{K}{T_1 s^2 + T_2 s + 1} = \frac{0.5}{0.5s^2 + 1.5s + 1} = \frac{1}{(s+1)(s+2)} \tag{4.72}$$

with the following performance specifications: percent of overshoot less than 4%, settling time $T_{set} = 1.2s$ and zero steady-state error for step input.

For the controller, it is advised to choose a minimum degree polynomials. Since zero steady-state error for step input is desired (the controller shall have one pole at the origin), we choose:

$$A(s) = l_2 s^2 + l_1 s \tag{4.73}$$

$$B(s) = k_2 s^2 + k_1 s + k_0 \tag{4.74}$$

Thus, the controller contains a pole at the origin.

In this case the target characteristic polynomial is written as:

$$q(s) = \frac{\tau^4}{\gamma_1^3 \gamma_2^2 \gamma_3} s^4 + \frac{\tau^3}{\gamma_1^2 \gamma_2} s^3 + \frac{\tau^2}{\gamma_1} s^2 + \tau s + 1 = Z_4 s^4 + Z_3 s^3 + Z_2 s^2 + Z_1 s + Z_0 \tag{4.75}$$

Since the settling time is given the equivalent time constant τ is:

$$\tau = T_{set}/2.5 = 1.2/2.5 = 0.48 \tag{4.76}$$

The stability indices are chosen according to the standard Manabe form:

$$\begin{aligned} \gamma_1 &= 2.5 \\ \gamma_2 &= 2 \\ \gamma_3 &= 2 \end{aligned} \tag{4.77}$$

The characteristic polynomial is:

$$q(s) = D(s)A(s) + N(s)B(s) \tag{4.78}$$

with $D(s) = T_1 s^2 + T_2 s + 1$, $A(s) = l_2 s^2 + l_1 s$, $N(s) = K$ and $B(s) = k_2 s^2 + k_1 s + k_0$, then we obtain:

$$q(s) = T_1 l_2 s^4 + (T_1 l_1 + (1 + 2T_1) l_2 - K k_2) s^3 + \\ +(2l_2 + (2T_1 + 1) l_1 - K k_1 + 2K k_2) s^2 + (2l_1 - K k_0 + 2K k_1) s + 2K k_0 \tag{4.79}$$

Using coefficient identification techniques according to 4.75 we obtain:

$$\begin{aligned} l_2 &= Z_4/T_1 \\ l_1 &= (Z_3 - l_2 T_2)/T_1 \\ k_2 &= (Z_2 - l_2 - l_1 T_2)/K \\ k_1 &= (Z_1 - l_1)/K \\ k_0 &= Z_0/K \end{aligned} \tag{4.80}$$

The controller is $K(s) = B(s)/A(s)$, where:

$$\begin{aligned} A(s) &= l_2 s^2 + l_1 s = 0.0008493 s^2 + 0.0151467 s \\ B(s) &= k_2 s^2 + k_1 s + k_0 = 0.1371813 s^2 + 0.9297066 s + 2 \end{aligned} \tag{4.81}$$

The prefilter (set-point filter) is given by:

$$\frac{F}{B(s)} = \frac{k_0}{B(s)} = \frac{2}{0.1371813 s^2 + 0.9297066 s + 2} \tag{4.82}$$

The step response of the system is shown in Fig. 4.18, the step input is applied at $t = 0.5s$ and the system has no overshoot.

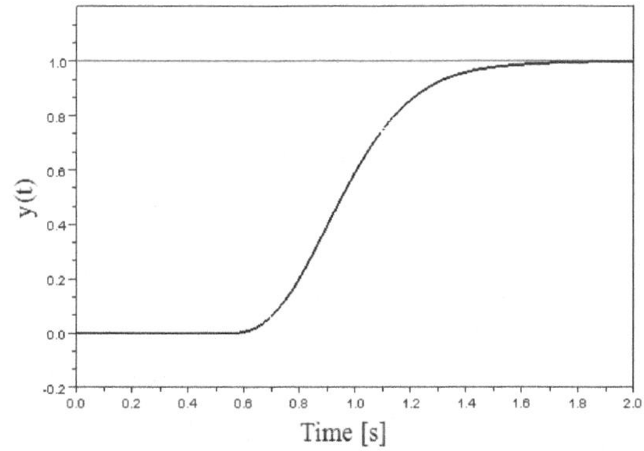

Fig. 4.18 Step response.

Finally we check the obtained gain and phase margins: $G_m = \infty$ and $\phi_m = 39.5°$.

We will return to this design method in the next chapter, where the same procedure will be exemplified in case of a system with time delay.

4.6 Validation of the Control System

The previous control designs have been performed only for the nominal plant, no plant uncertainty has been taken into account. In order to assure robust performance, the control system must be designed for a set of plants, defined by the nominal plant and plant uncertainty.

The system's parameters might undergo variations due to:

- temperature variation
- changing in the operating point
- manufacturing dispersion
- natural aging of components.

The obvious question is: what is the range of parameter variations, which shall be considered during control design? In the aim to answer this question we introduce the concept of representative sample and sample size.

Having defined the range of the parameters, we can validate the control system using Monte Carlo simulation. If the control systems fulfills all the specifications - when the parameters undergo variations - the design can be accepted, otherwise we have to iterate on it.

4.6.1 Representative Sample and Sample Size

In order to find out the range of parameters, we have to take some samples (random sampling is recommended). If the sample size increase then the precision and confidence level becomes higher.

We can determine the appropriate size of the representative sample. Therefore, we have to specify some measure of precision that is required to be obtained. Precision measures are specified in terms of:

- confidence intervals;
- relative precision of mean and variance (let us note the relative precision with r_p).

Usually we require that the confidence interval to be 90% or 95%. Precision essentially depends only upon the sample size, not the population size (in manufacturing, population size basically means batch size).

For example, suppose that we wish to be 95% sure (confidence interval 95%) that the sample mean is within 10% relative precision of the true mean ($r_p = 0.1$). Then using the formula derived in [75] - the main results are summarized in Table 4.6 - we can easily compute the sample size n:

$$n = \frac{TH_c}{r_p^2} = \frac{1.2736}{0.01} \approx 128 \tag{4.83}$$

where TH_c is the Thompson coefficient and r_p is the relative precision.

Table 4.6 Sample size based on Thompson formula.

Confidence interval	TH_c Thompson coeff.	Sample size for r_p =0.1	Sample size for r_p =0.05	Sample size for r_p =0.02
90%	1.0064	101	403	2516
95%	1.2736	128	510	3184
97.5%	1.5596	156	624	3899
98%	1.6587	166	664	4147
99%	1.9699	197	788	4925

There are also other approaches to compute the sample size, considering previous values of mean and variance (based on previous results or pilot studies), but the presented method has the main advantage of simplicity.

It is important to remark, that there is a dramatic rise in the value of n (sample size) as the confidence interval increases and the relative precision r_p goes down from 10% (0.01) to 2% (0.02) as shown in Table 4.6.

After the size of the representative sample is defined based on measurements we can easily compute the mean and the variance of each parameter.

We recall that the variance λ is the second central moment:

$$\lambda = \sigma^2 \tag{4.84}$$

The standard deviation σ is the positive square root of the variance λ. The standard deviation σ has a nice interpretation in the case of a Gaussian distribution:

$$\begin{aligned} P(-\sigma < e < \sigma) &= 0.682 \\ P(-2\sigma < e < 2\sigma) &= 0.954 \\ P(-3\sigma < e < 3\sigma) &= 0.997 \end{aligned} \tag{4.85}$$

That means that most of the values (99.7%) are in the interval $-3\sigma < e < 3\sigma$ or -$3\sigma + \mu < e < 3\sigma + \mu$, where μ is the mean value.

Let assume that we would like to know the mean and the variance of a coil parameters: resistance R and inductance L, having a confidence interval of 95% and relative precision 10%. Now considering a representative sample having 128 elements we can compute the mean and the standard deviation of the parameters. Let's say that we have obtained:

$$\begin{aligned} R_\mu &= 6.8\Omega \; R_\sigma = 0.05\Omega \\ L_\mu &= 0.5H \; L_\sigma = 0.03H \end{aligned} \tag{4.86}$$

The next step is to simulate our system assuming a normal distribution of the parameter having mean and variance given in equation (4.86). This approach is called Monte Carlo simulation and is briefly presented in the next section. As a remark, for simplicity and better understanding we are going to simulate the open-loop system, but the procedure is similar for closed-loop systems.

4.6.2 Monte Carlo Simulation

Having the parameter variation range in terms of mean and variance as well as considering a normal distribution we can test the robustness and reliability of the designed control system using so called Monte Carlo simulation.

The main steps of the Monte Carlo simulation are as follows:

- for each uncertain variable (one that has a range of possible values), define the possible values with (let's say a normal) probability distribution based on the values obtained from the representative sample
- generate the system's parameters in a random way according to their probability distributions
- simulate the control system for the generated parameter data set
- assess the control system performance and robustness using the simulation results.

Example 4.7. *Perform Monte Carlo simulation for the system (a coil) given by the following transfer function:*

$$P(s) = \frac{I(s)}{V(s)} = \frac{1}{sL + R} \tag{4.87}$$

where v is the applied voltage and i is the current and the parameters are defined according to equation (4.86).

The Matlab's *randn* function generates arrays of random numbers whose elements are normally distributed with 0 mean and variance 1. Using this Matlab function we generate the random resistance and inductance with mean and variance given by equation (4.86). The generated random parameters are presented in Fig. 4.19 and Fig. 4.20.

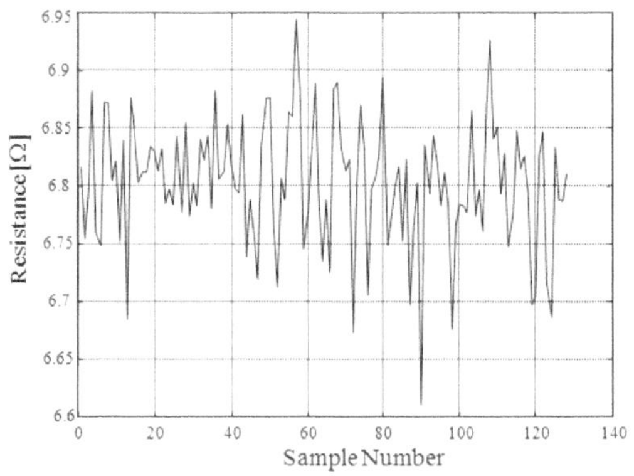

Fig. 4.19 Generated random resistance.

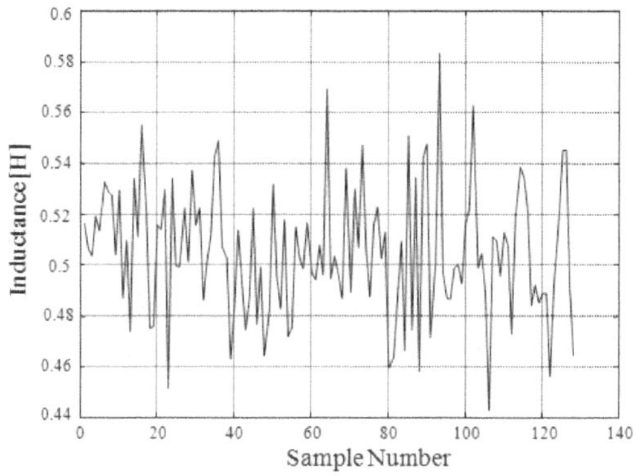

Fig. 4.20 Generated random inductance.

The system response, when the parameters undergo variations is presented in Fig. 4.21, when the applied input voltage is a step with $v = 48V$.

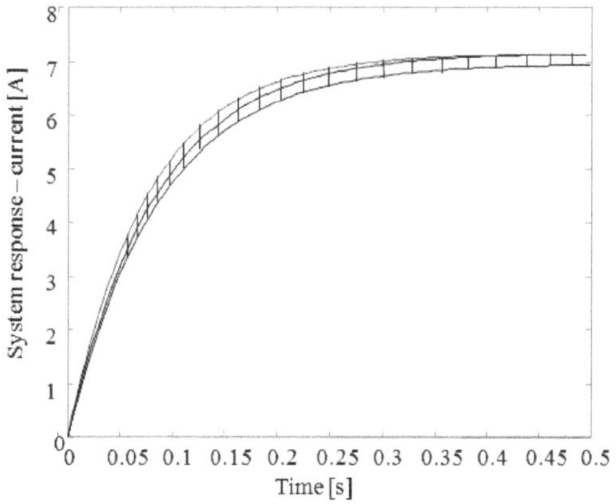

Fig. 4.21 The system response for random parameters.

Main Steps of Monte Carlo Simulation Applied for Control Systems

- define the confidence interval and relative precision of mean and variance
- using Thompson's formula compute the sample size
- based on measured data compute the mean and the variance of parameters
- generate the system's parameters in random way according to their probability distributions
- simulate the system for the generated data set

- assess the system performance (i.e. settling time, percent of overshoot, etc.), robustness (gain and phase margins) as well as reliability (i.e. sensor fails and the closed-loop system opens) using simulation results.

Monte Carlo simulation can be viewed as a simple and useful method to assess the system performance and robustness to parameter variations. It is recommended to perform such simulation considering at least:

- parameter variations due to temperature variation
- parameter variation due to manufacturing dispersion
- parameter variation due to components aging.

Chapter 5

Classical Controller Design - Part II

"Memories are what warm you up from the inside. But they're also what tear you apart." — Haruki Murakami, Kafka on the Shore

Summary:

This chapter is dedicated to controller design in a classical framework. The following items are discussed:

- controller design for systems with time delays using the Smith predictor as well as the coefficient diagram method
- controller design for disturbance rejection as well as notch filters
- application of back-EMF observers
- two-degree-of-freedom 2DOF control systems.

5.1 Controller Design for Systems with Time Delays

In this section we are going to discuss controller design techniques for systems with time delays. In the industry - especially process industry - there are many systems with time delays, which have to be controlled. Furthermore, in the past decades control and distributed control over communication networks evolved significantly and are widely used.

The general structure of an embedded control system over a communication network is shown in Fig. 5.1.

The main components of the system are: the plant, which is going to be controlled, the sensor that contains analog-to-digital converter and communication interface, the embedded system on which the control algorithm is implemented, the actuator which contains a communication interface and a digital-to-analog converter and finally the communication network.

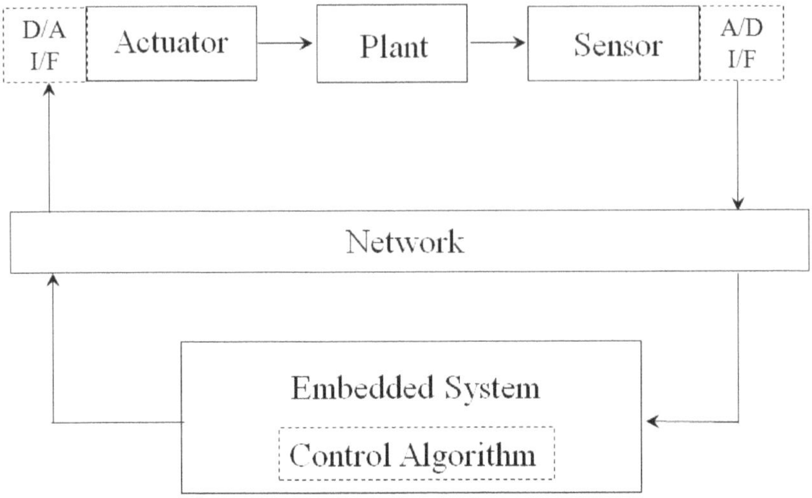

Fig. 5.1 Embedded control system over a communication network.

The typical situation is that the plant shall be controlled under varying network delay. There are also other interesting topics such as: control under data loss, control under communication bandwidth limitation, control under network topology constraints, but they are out of scope of this book.

Another communication topology, frequently found in the industry as well as in vehicles is a distributed control system over a communication network, shown in Fig. 5.2.

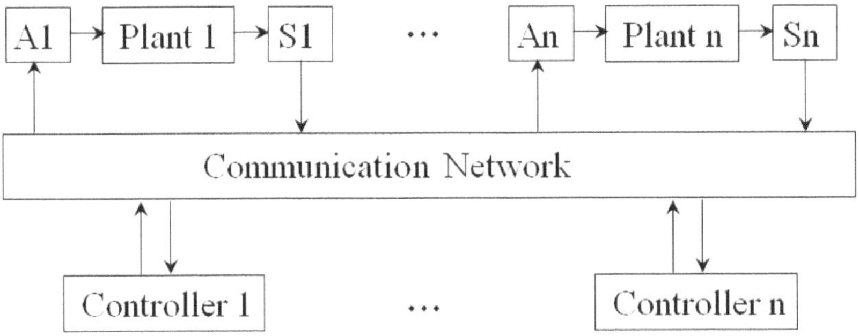

Fig. 5.2 Distributed control system over a communication network.

Control systems over networks introduce inherent time delays, which have to be handled by the control algorithm. Therefore, we will present simple control design techniques for systems with time delays.

5.1.1 Systems with Time Delays – Smith Predictor

In the 50-ties Smith suggested a control algorithm, which could be applied effectively to control systems with significant dead time. The main idea is that the feedback control system is transformed into a control system where the dead time is outside of the feedback circuit.

The two control diagrams shown in Fig. 5.3 are equivalent. This can be proved if we write down the transfer functions of the two systems.

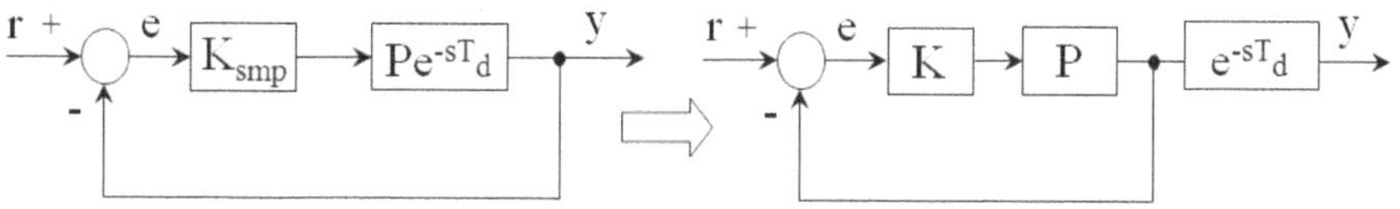

Fig. 5.3 Equivalent block diagrams - Smith predictor.

$$\frac{K_{smp}Pe^{-sT_d}}{1+K_{smp}Pe^{-sT_d}} = \frac{KPe^{-sT_d}}{1+KP} \tag{5.1}$$

$$K_{smp} = \frac{K}{1+KP(1-e^{-sT_d})} \tag{5.2}$$

where the term e^{-sT_d} is the transfer function of the pure delay element.

The Smith predictor employs a model of the plant characterized by a linear transfer function and a time delay. It uses a model to simulate both the delayed and undelayed states of the plant as shown in Fig. 5.4. Although the Smith predictor has the capability of transforming a time-delay control design to a delay free problem, there is a controversial issue with regard to its sensitivity to plant parameter variations.

The advantage of this approach is that the controller is designed independently of time delay.

Example 5.1. *Let us consider the following control problem, where the plant with time delay $T_d = 0.1s$ is:*

$$P(s) = \frac{1}{(s+1)(s+2)}e^{-sT_d} \tag{5.3}$$

If we use the deadbeat compensator designed in the previous section for the plant without time delay:

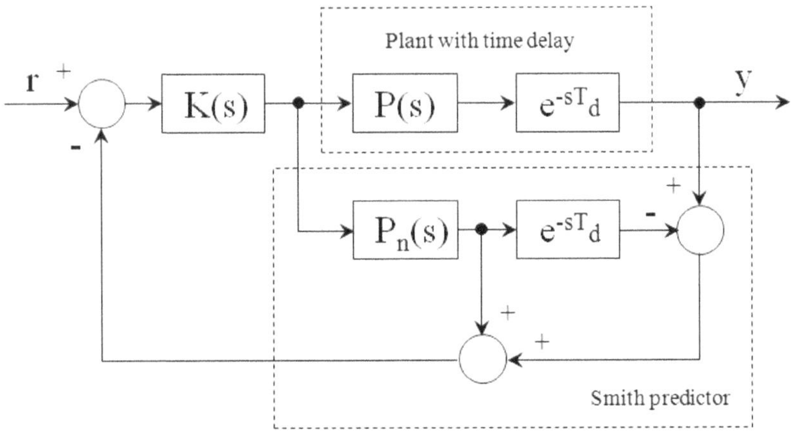

Fig. 5.4 Control of systems with time delay using Smith predictor.

$$K(s) = k\frac{(s+z_1)(s+z_2)}{s(s+p)} = 36.6\frac{s^2+4.58s+7}{s(s+5.8)} \tag{5.4}$$

with the prefilter:

$$K_p(s) = \frac{z_1 z_2}{(s+z_1)(s+z_2)} = \frac{7}{s^2+4.58s+7} \tag{5.5}$$

the output of the control system is oscillating (shown in Fig. 5.5 dashed line). If we compensate the dead time of the system using a Smith predictor the system behaviour can be improved significantly, shown in Fig. 5.5 solid line. Finally, we check the obtained gain and phase margins: $G_m = 7$ and $\phi_m = 16°$.

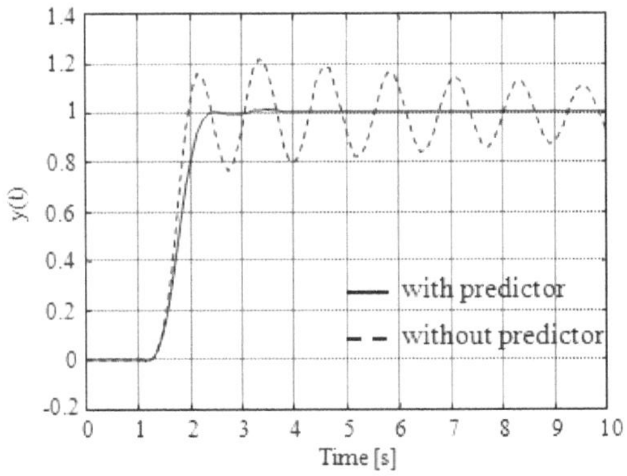

Fig. 5.5 Step response - Smith predictor.

The Smith predictor can be considered an effective way to control systems with constant time delays, but the control performance is highly influenced by the model accuracy.

5.1.2 The Coefficient Diagram Method (CDM) for Systems with Time Delays

In the previous chapter, we presented the the coefficient diagram method, now we will exemplify, how to use it in case of systems with time delays.

Let us consider a first-order plant with time delay and a second-order plant with time delay described as:

$$P_1(s) = \frac{K}{Ts+1}e^{-sT_d} \tag{5.6}$$

$$P_2(s) = \frac{K}{T_2s^2+T_1s+1}e^{-sT_d} \tag{5.7}$$

The term e^{-sT_d} represents the time delay and is approximated by a first-order Padé approximation.

$$e^{-sT_d} \approx \frac{2-T_ds}{2+T_ds} \tag{5.8}$$

The first-order Padé approximation is sufficient in may cases and has the advantage that the resulting equivalent plant as well as controller have low-order.

We will note with:

$$P_{eq1}(s) = \frac{-KT_ds+2K}{TT_ds^2+(2T+T_d)+2} \tag{5.9}$$

$$P_{eq2}(s) = \frac{-KT_ds+2K}{T_1T_ds^3+(2T_1+T_2T_d)s^2+(2T_2+T_d)s+2} = \frac{N(s)}{D(s)} \tag{5.10}$$

the approximated plants.

The block diagram of the control structure is show in Fig. 5.6.

For the first-order plant with time delay we will consider the controller polynomials as:

$$A(s) = l_2s^2 + l_1s \tag{5.11}$$

$$B(s) = k_2s^2 + k_1s + k_0 \tag{5.12}$$

For the second-order plant with time we will consider the controller polynomials as:

$$A(s) = l_3s^3 + l_2s^2 + l_1s \tag{5.13}$$

$$B(s) = k_3s^3 + k_2s^2 + k_1s + k_0 \tag{5.14}$$

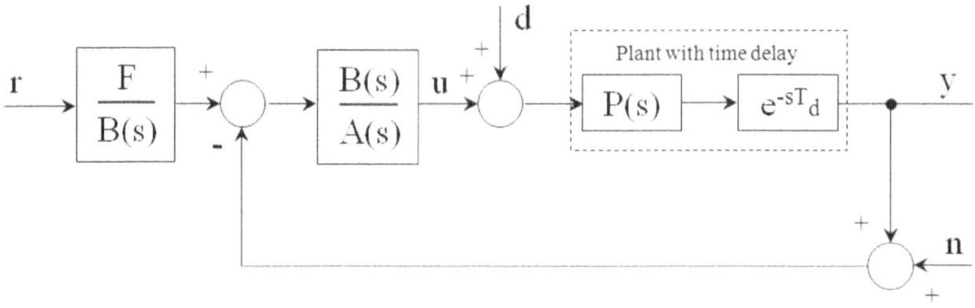

Fig. 5.6 Coefficient diagram control structure for plants with time delays.

We will illustrate how the controller parameters are computed only in case of the second-order plant with time delay. However, at the end of this section the controller parameters for the first-order system with time delay are also given.

The characteristic polynomial is written as:

$$q(s) = A(s)D(s) + B(s)N(s) \tag{5.15}$$

In this case, the target characteristic polynomial is written as:

$$q(s) = \frac{\tau^6}{\gamma_1^5\gamma_2^4\gamma_3^3\gamma_4^2\gamma_5}s^6 + \frac{\tau^5}{\gamma_1^4\gamma_2^3\gamma_3^2\gamma_4}s^5 + \frac{\tau^4}{\gamma_1^3\gamma_2^2\gamma_3}s^4 + \frac{\tau^3}{\gamma_1^2\gamma_2}s^3 + \frac{\tau^2}{\gamma_1}s^2 + \tau s + 1 \tag{5.16}$$

or in a more compact form as:

$$q(s) = Z_6s^6 + Z_5s^5 + Z_4s^4 + Z_3s^3 + Z_2s^2 + Z_1s + Z_0 \tag{5.17}$$

The terms $Z_6, ..., Z_0$ are known, since the equivalent time constant τ and the stability indices γ_i for $i = 1, ...5$ are design parameters. Using coefficient identification we obtain seven linear algebraic equations having seven unknowns. Solving the equations the following explicit formulae are found [40]:

$$l_3 = \frac{Z_6}{T_1T_d} \tag{5.18}$$

$$l_2 = \frac{Z_5 - (2T_1 + T_2T_d)l_3}{T_1T_d} \tag{5.19}$$

$$W = Z_4 - (2T_2 + T_d)l_3 - (2T_1 + T_2T_d)l_2 \tag{5.20}$$

$$V = Z_3 - 2l_3 - (2T_2 + T_d)l_2 \tag{5.21}$$

$$X = Z_2 - 2l_2 \tag{5.22}$$

$$l_1 = \frac{T_d^4 + 2\tau T_d^3 + 4XT_d^2 + 8VT_d + 16W}{8T_d(T_d^2 + 2T_2T_d + 4T_1)} \tag{5.23}$$

$$k_3 = \frac{T_dT_1l_1 - W}{KT_d} \tag{5.24}$$

$$k_2 = \frac{2Kk_3 + (2T_1 + T_dT_2)l_1 - V}{KT_d} \tag{5.25}$$

$$k_1 = \frac{\tau + 0.5T_d - 2l_1}{2K} \tag{5.26}$$

$$k_0 = 0.5/K \tag{5.27}$$

Finally, the controller $K(s)$ and the prefilter is obtained as:

$$K(s) = \frac{B(s)}{A(s)} \tag{5.28}$$

$$\frac{F}{B(s)} = \frac{k_0}{B(s)} \tag{5.29}$$

Example 5.2. *Let us consider the following control problem, where the plant with time delay $T_d = 0.1s$ is:*

$$P(s) = \frac{1}{(s+1)(s+2)}e^{-sT_d} \tag{5.30}$$

The performance specifications are: percent of overshoot less than 4%, settling time $T_{set} = 1.2s$ and zero steady-state error for step input.

For this plant we have:

$$K = 0.5\ T_1 = 0.5\ T_2 = 1.5\ T_d = 0.1. \tag{5.31}$$

The design parameters are:

$$\begin{aligned} &\tau = T_{set}/2.5 \\ &\gamma_1 = 2.5 \\ &\gamma_2 = \gamma_3 = \gamma_4 = \gamma_5 = 2 \end{aligned} \tag{5.32}$$

Solving the equations we obtain the polynomials, which describe the controller and the prefilter:

$$\begin{aligned} &A(s) = 0.0000024s^3 + 0.0001476s^2 + 0.0083336s \\ &B(s) = 0.0033862s^3 + 0.0916973s^2 + 0.5133328s + 1 \\ &F = 1 \end{aligned} \tag{5.33}$$

The coefficients of polynomials $A(s)$ and $B(s)$ are much smaller than 1, which might create the impression that it will be difficult to implement in real-time, but by proper scaling implementation is possible.

Finally, we check the obtained gain and phase margins: $G_m = 4.95$ and $\phi_m = 31.6°$.

The designer has the choice to change the stability indices. The most effective indices on stability and overshoot are γ_1 and γ_2. Furthermore, in order to improve the phase and gain margin, we can choose:

$$\begin{aligned} \gamma_1 &= 3 \\ \gamma_2 &= 1.1 \\ \gamma_3 &= 4.5 \end{aligned} \tag{5.34}$$

The other γ values are unchanged. Next, we compute the controller paramters and we check the obtained gain and phase margins: $G_m = 6$ and $\phi_m = 32.41°$.

We remark, this is not the standard Manabe form, so the so the relation between the equivalent time constant τ and settling time does no hold exactly.

In case of a first-order system with time delay, described by equation 5.9 the coefficients of the controller polynomials $A(s)$ and $B(s)$ are:

$$l_2 = \frac{Z_4}{TT_d} \tag{5.35}$$

$$l_1 = \frac{T_d^3 + 2Z_1T_d^2 + 4(Z_2 - 4l_2)T_d + 8(Z_3 - 2Tl_2)}{8T_d(2T + T_d)} \tag{5.36}$$

$$k_2 = \frac{T_dTl_1 + (2T + T_d)l_2 - Z_3}{KT_d} \tag{5.37}$$

$$k_1 = \frac{\tau + 0.5T_d - 2l_1}{2K} \tag{5.38}$$

$$k_0 = 0.5/K \tag{5.39}$$

It is left for the reader to design a controller using CDM for the first-order plant with time delay $T_d = 0.1s$, given below:

$$P(s) = \frac{1}{s+2}e^{-sT_d} \tag{5.40}$$

The performance specifications are: percent of overshoot less than 4%, settling time $T_{set} = 1.2s$ and zero steady-state error for step input.

Using the standard Manabe form and solving the equations we should obtain the following polynomials:

$$\begin{aligned} A(s) &= 0.0084935s^2 + 0.0417109s \\ B(s) &= 0.0516200s^2 + 0.4465782s + 1 \\ F &= 1 \end{aligned} \tag{5.41}$$

Finally, we check the obtained gain and phase margins: $G_m = 9.41$ and $\phi_m = 39.8°$.

Remark 5.1. *In control applications, time delay corresponds to a phase advance, which generally reduces stability margins.*

5.1.3 Handling Jitter in Networked Control System

Till now we assumed that the time delay is constant. Especially in networked control systems, the time delay might vary and this effect is know as communication jitter.

If the jitter cannot be reduced, the control engineer shall find a solution to minimize the effect of the jitter on the control system.

To illustrate the situation, let us assume a position sensor equipped with CAN (Control Area Network) communication interface. The sensor is sampling the position at every 2.5 ms (milliseconds) but sends the CAN message containing the position information only at every 25 ms.

The CAN message should be received at every 25 ms, but measurements presented in Fig. 5.7 show that there is communication jitter, which exceed even 10% of the message rate.

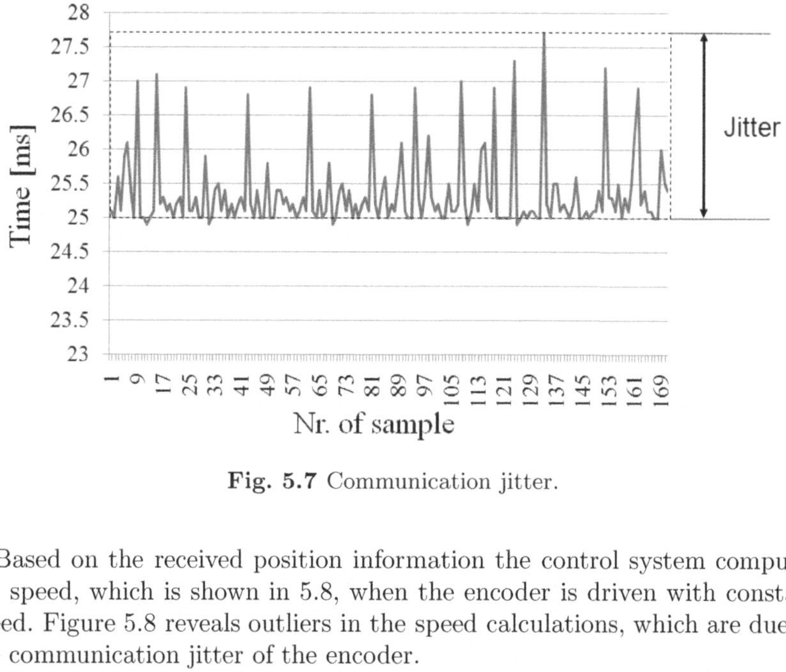

Fig. 5.7 Communication jitter.

Based on the received position information the control system computes the speed, which is shown in 5.8, when the encoder is driven with constant speed. Figure 5.8 reveals outliers in the speed calculations, which are due to the communication jitter of the encoder.

The simplest approach to improve the precision of the speed information is to filter online the outliers. If linear filter is used in presence of outliers is likely that poor prediction is obtained. Therefore a simple data cleaning (nonlinear) filter is presented.

Let us note with y_k the measured data sequence, x_k is the "nominal" data sequence in which we are interested and o_k represents a sequence of contaminating outliers.

We also assume that the outlier can be described by the additive outlier model [59]:

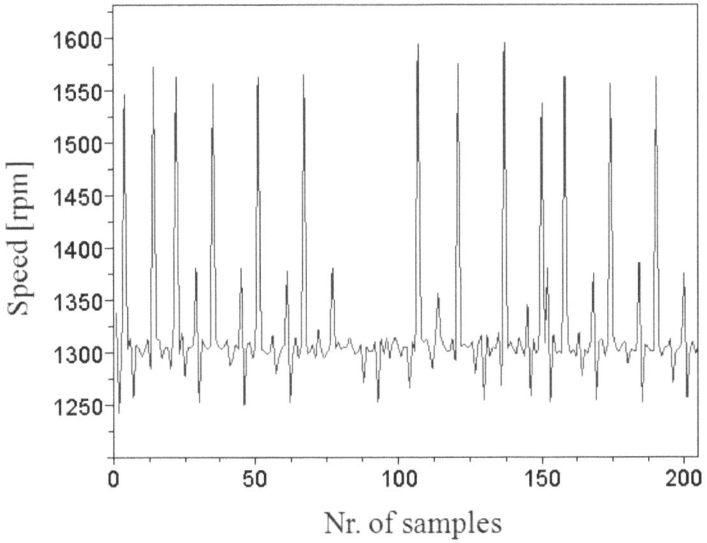

Fig. 5.8 Calculated speed in presence of communication jitter.

$$y_k = x_k + o_k \tag{5.42}$$

We seek an approximation of x_k based only on the current and past data observations. We define a data window of width N as:

$$W_k = \{y_k, y_{k-1}, ..., y_{k-N+1}\} \tag{5.43}$$

Then we order the values in this window in ascending order as:

$$R_k = y^k_{(1)} \leq y^k_{(2)} \leq ... \leq y^k_{(N)} \tag{5.44}$$

and the median y_k^{med} of the sequence R_k is computed as:

$$y_k^{med} = \begin{cases} y^k_{((N+1)/2)} & for\ N\ \ odd \\ (y^k_{(N/2)} + y^k_{(N/2+1)})/2 & for\ N\ even \end{cases} \tag{5.45}$$

Next, we define the absolute distance between the current value and the median value:

$$d_k = |y_k^{med} - y_k| \tag{5.46}$$

If the distance exceeds some specified threshold $TH \geq 0$, we declare y_k to be an outlier and replace it with a prediction y_k^* and we obtain the filtered sequence f_k:

$$f_k = \begin{cases} y_k\ if\ d_k \leq TH \\ y_k^*\ if\ d_k > TH \end{cases} \tag{5.47}$$

The data cleaning filter requires the following specifications:

- the value y_k^* with which the outlier is replaced
- the threshold value TH, which we can even make time dependent
- the window width N
- filter initialization, how f_k is defined for $k < N$.

One of the simplest possibilities is the median replacement for which:

$$y_k^* = y_k^{med} \tag{5.48}$$

Figure 5.9 shows the speed (dashed line) and the median filtered speed (solid line) considering: the threshold value $TH = 157[rpm]$, the window length $N = 4$ and filter initialization is $f_k = 0$ for $k < 4$

The described simple outlier filter is based on a moving data window that is suitable for real-time applications like closed loop control. If the current data value y_k lies far from the median value - in the data window - is declared outlier and is replaced by the median value. Otherwise, the filter leaves y_k unmodified.

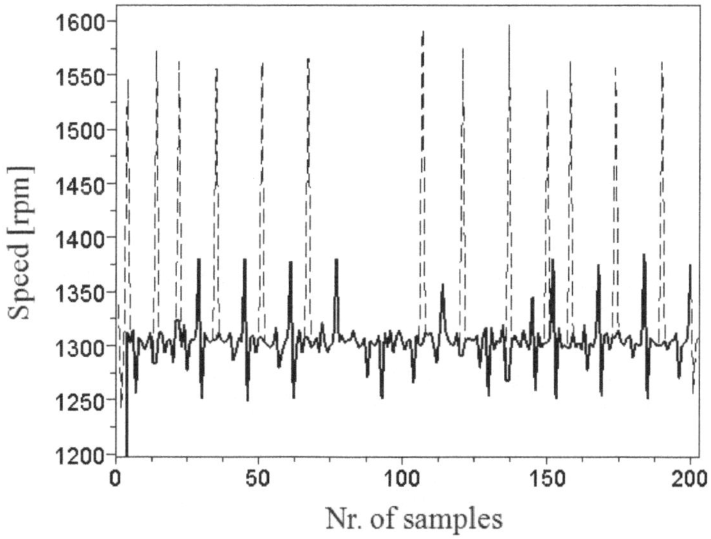

Fig. 5.9 Speed and median filtered speed.

5.2 Controller Design for Disturbance Rejection

This section deals with a disturbance rejection control problem in the case of a lightly damped structure. The simplest way to increase the damping of a lightly damped structure is to use so-called rate feedback.

Let assume that the plant transfer function is given by:

$$P(s) = 1.5\frac{20^2}{s^2 + 0.2s + 20^2} \tag{5.49}$$

If we use a rate feedback loop $K(s) = ks$ - as illustrated in Fig. 5.10 - the closed-loop transfer function from the external disturbance to the control system output becomes:

$$T_{yd}(s) = \frac{P(s)}{1 + P(s)K(s)} = 1.5\frac{20^2}{s^2 + (0.2 + 1.5 \cdot 20^2 \cdot k)s + 20^2} \tag{5.50}$$

It is easy to remark that the damping ratio is increased by the rate feedback. This approach is very simple, however might amplify excessively the unmodelled resonance peaks at high frequencies.

Therefore, for SISO systems, a slow roll-up lead compensator can be designed. The compensator has the following structure:

$$K(s) = k\frac{(s+m)(s+4m)(s+16m)...}{(s+2m)(s+8m)(s+32m)...} \tag{5.51}$$

where k and m are design parameters. Its gain is rolling up at an average slop of 10dB/decade to avoid over-amplifying resonance peaks at high frequencies.

Next, let us consider a similar plant given by:

$$P(s) = 1.5\frac{20^2}{s^2 + 0.2s + 20^2} + 1.3\frac{70^2}{s^2 + s + 70^2} \tag{5.52}$$

The Bode plot of $P(s)$ is represented by the solid line in Fig. 5.11. If we choose a slow roll-up compensator having the form:

$$K(s) = k\frac{(s+m)(s+4m)(s+16m)}{(s+2m)(s+8m)(s+32m)} \tag{5.53}$$

with $k = 0.3$ and $m = 5$, then is possible to damp the lightly damped poles, as shown in Fig. 5.11 by the dotted line.

5.2.1 Notch Filters

There are practical applications when the transfer function of the controlled plant contains one or more pairs of complex-conjugate poles that lie close to the imaginary axis in the s-plane (lightly damped poles).

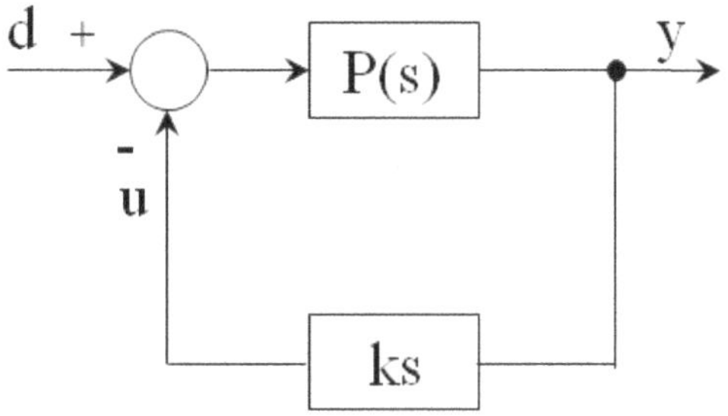

Fig. 5.10 Rate feedback control structure.

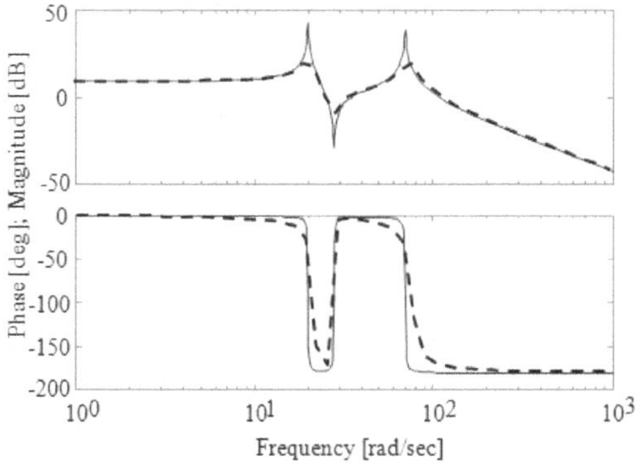

Fig. 5.11 Damped and undamped system.

Example 5.3. *Let us consider the following plant:*

$$P(s) = \frac{1}{(s+1)(s+2)} \cdot \frac{120}{s^2 + 5s + 120} \tag{5.54}$$

Basically, the plant considered in the previous examples has been augmented with two lightly damped poles.

One way to control this system is to design a controller with zeros, near the undesirable, lightly-damped poles of the plant. These zeros can attenuate the effect of these poles. Such controller is called notch filter. The block diagram of the control system with notch filter is shown in Fig. 5.12.

It should be noted that due to the nature of most systems, exact pole/zero cancellation cannot be obtained, nor should be attempted, approximate cancellation will give us many of the desirable characteristics.

For the example above, let's try placing the zeros slightly to the left of the lightly damped poles - it is a good idea to pull the poles to the left instead of the right.

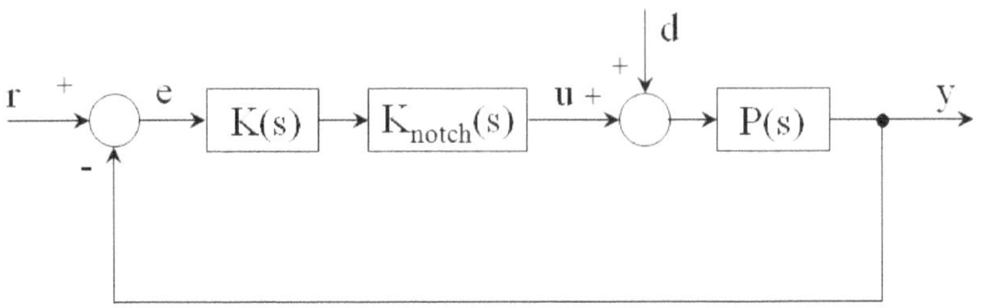

Fig. 5.12 Control system with notch filter.

Let us try the following compensator (notch filter):

$$K_{notch}(s) = 20\frac{s^2 + 7s + 120}{s^2 + 100s + 2400} \tag{5.55}$$

in combination with the deadbeat controller:

$$K(s) = 36.6\frac{s^2 + 4.58s + 7}{s(s + 5.8)} \tag{5.56}$$

The system response in the case of a step input is shown in Fig. 5.13 and the obtained gain and phase margins are: $G_m = 8.03$ and $\phi_m = 23.78°$.

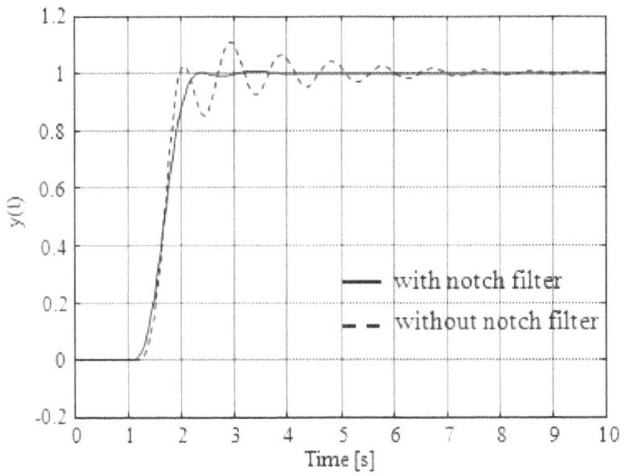

Fig. 5.13 Step response - the influence of notch filter.

5.3 Disturbance Observers

A disturbance observer (DOB) is a simple and useful approach to handle disturbances and other uncertainties in the feedback system [33]. The principle of the DOB is shown in Fig. 5.14 where $P(s)$ is the real plant, $P_n(s)$ is the nominal plant model, u is the plant input, y is the plant output, d is the external disturbance and d^* is the estimated disturbance.

In this approach, the actual output of the controlled plant y is processed by an approximate realizable inverse of the nominal dynamics of the nominal plant $1/P_n(s)$, and the difference between the processed signal and the control input is regarded as the equivalent disturbance d^*. As a result, disturbances and model uncertainties in a specified range are removed. The overall transfer function of this block diagram is described as follows, where $F(s)$ is the filter transfer function.

$$Y(s) = G_{YU}(s)U(s) + G_{YD}(s)D(s) \tag{5.57}$$

$$G_{YU} = \frac{PP_n}{F(P - P_n) + P_n} \tag{5.58}$$

$$G_{YD} = \frac{PP_n(1 - F)}{F(P - P_n) + P_n} \tag{5.59}$$

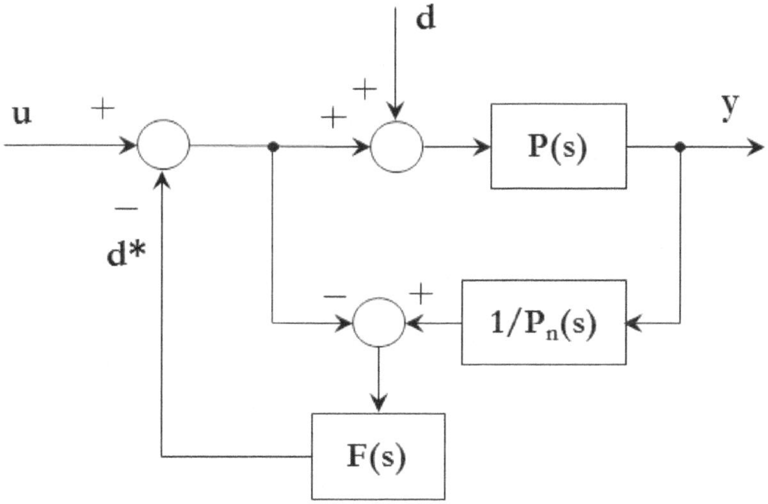

Fig. 5.14 Principle of disturbance observer.

At the frequencies, where $F(s)$ is close to 1, the frequency response G_{YD} is close to 0 (i.e. the disturbance is completely reduced). G_{YU} approximates $P_n(s)$, which indicates that the input-output character of this system behaves like the nominal plant. This shows that $F(s)$ should be designed as a low pass filter. The block diagram shown in Fig. 5.14 cannot be realized because $1/P_n(s)$ is normally unrealisable. Therefore, it is transformed into the equivalent block diagram shown in Fig. 5.15 so that the structure can be implemented. In this case $F(s)/P_n(s)$ has to be realizable.

As a remark, a higher cut-off frequency of the $F(s)$ filter, will lead to improved disturbance rejection performance and wider disturbance rejection range, but the stability margins is reduced.

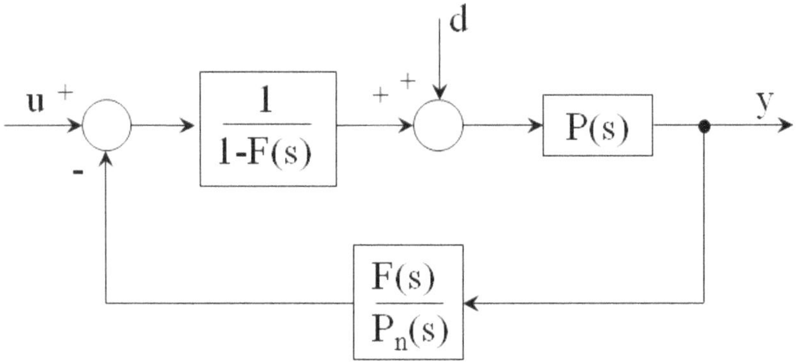

Fig. 5.15 Implementation of disturbance observer.

Example 5.4. *Let us consider the following plant:*

$$P(s) = \frac{1}{(s+1)(s+2)} \tag{5.60}$$

and let us assume that at the moment $t = 3s$ *an external disturbance is applied as a step input* $d(t) = 10$.

Let us consider the following disturbance observer with:

$$F(s) = \frac{\omega_d^2}{s^2 + 2\xi_d \omega_d s + \omega_d^2} \tag{5.61}$$

with $\xi_d = 0.8$ and $\omega_d = 200 rad/sec$.

Using the same deadbeat controller with the known prefilter the system response is shown in Fig. 5.16. The dotted line marks the system response

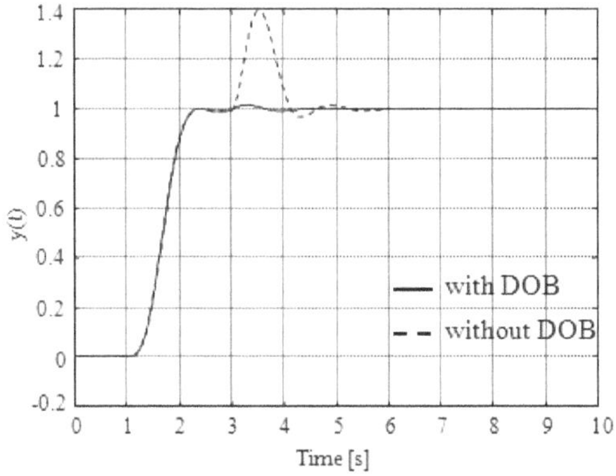

Fig. 5.16 Step response - influence of DOB.

without disturbance observer and the solid line marks the system response with disturbance observer.

It is easy to remark that the system response has been improved significantly due to the presence of disturbance observer.

5.3.1 Electromagnetic Armature Movement Detection

The basic structure of the electromagnetic actuator is presented in Fig. 5.17.

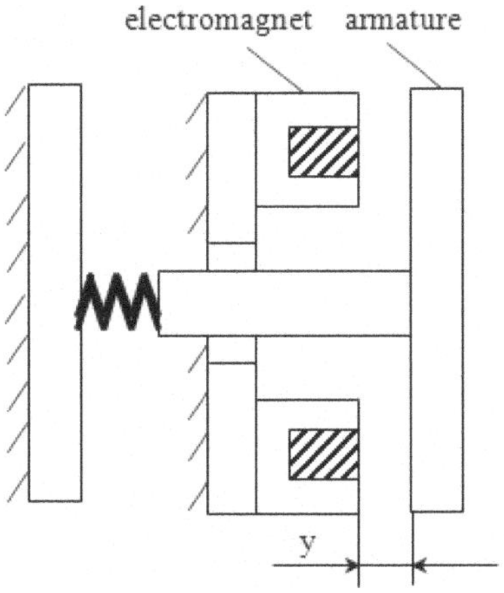

Fig. 5.17 The electromagnetic actuator.

The related mathematical model can be written as:

$$u(t) = Ri(t) + N\frac{d\phi}{dt} = Ri(t) + N\frac{\partial\phi}{\partial i}\frac{di}{dt} + N\frac{\partial\phi}{\partial y}\frac{dy}{dt} \tag{5.62}$$

where: u - applied voltage, i - current, R - coil resistance, N - number of coil turn and ϕ is the flux linked by the coil turns, $\phi = \phi(i, y)$, which is current i and air-gap y dependent.

The force generated by the electromagnet can be modelled by the nonlinear relationship:

$$F_m = k_f \phi_g^2 \tag{5.63}$$

where ϕ_g refers to the magnetic gap flux and k_f is a constant dependent upon actuator geometry.

The differential equation of motion is:

$$m\frac{d^2y}{dt^2} = F_s - F_m \tag{5.64}$$

where m is the moving mass and F_s is the spring force and F_m is the force generated by the electromagnet.

If the actuator is not saturated then $\phi = \phi(y)$ and the previous equation applying the Laplace transform becomes:

$$u = Ri + sLi + e \tag{5.65}$$

where L is the coil inductance and e is the induced electromotive force (EMF) or back-EMF.

Basically, there is a nonlinear dependence between the back-EMF an armature speed dy/dt. Since, back-EMF is induced by the movement if the back-EMF exceeds a defined threshold armature movement can be detected.

The back-EMF observer is given in equation 5.66 and is presented in Fig. 5.18, which resembles a disturbance observer.

$$\hat{e} = \frac{u - (Ls + R)i}{\tau s + 1} \tag{5.66}$$

where τ is the time constant of the low pass filter.

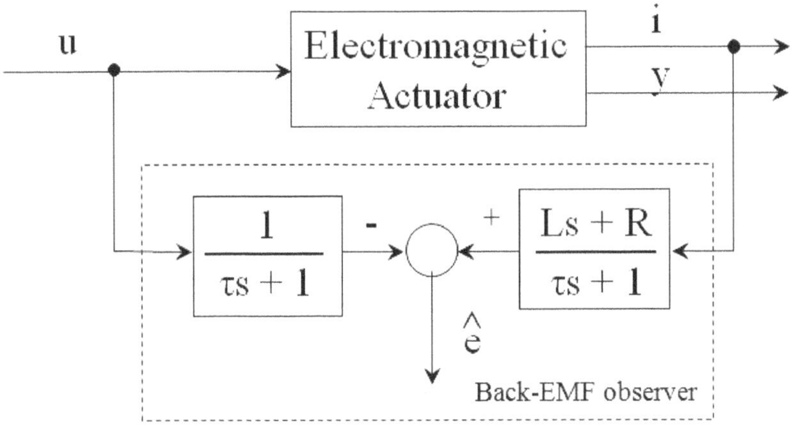

Fig. 5.18 Back-EMF observer for the electromagnetic actuator.

5.4 Two-Degree-of-Freedom Control Systems (2DOF)

The control systems having error feedback only are called one-degree-of-freedom (abbreviated 1DOF) control systems. There are significant advantages in having control systems with other configurations. An example of such system is shown in Fig. 5.19, which is called a system with two-degree-of-freedom (abbreviated 2DOF) because the signal paths from the control set point to the control signal is different from the signal path from the output to the control signal. This terminology was introduced by Horowitz who analysed these systems carefully [42].

A very nice property of systems with two-degree-of-freedom is that the problem of set point response can be separated from the problems of robustness and disturbance rejection.

In the case of a 2DOF control structure, we will first design a feedback by compromising between disturbance attenuation and robustness, when this is done we will then design a model and a feedforward, which gives the desired response to the set point [7].

By definition the degree of freedom of a control system is the number of closed-loop transfer functions that can be adjusted independently [7].

The general form of the two-degree-of freedom control system is presented in Fig. 5.19, where the compensator consists of two compensators $K(s)$ and $K_f(s)$. The $K(s)$ compensator is called serial compensator and the $K_f(s)$ controller is called feedforward compensator. The closed-loop transfer functions from r to y and d to y are:

$$G_{YR2} = \frac{P(K + K_f)}{1 + PKF} \tag{5.67}$$

$$G_{YD2} = \frac{P_d}{1 + PKF} \tag{5.68}$$

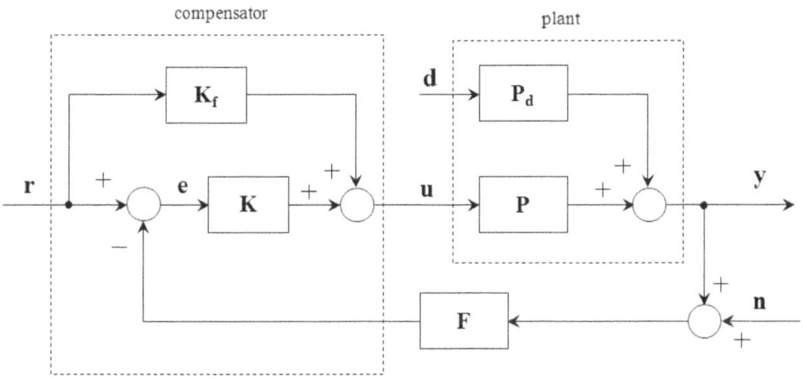

Fig. 5.19 Two-degree-of-freedom (2DOF) control system.

The subscript '2' means that the quantities are of the 2DOF control system.

It can be shown that the steady-state error to the unit step change of the set-point variable ϵ_r and the steady-state error to the unit step disturbance ϵ_d, become zero if:

$$\lim_{s\to 0} K(s) = \infty \; \lim_{s\to 0} K_f(s)/K(s) = 0 \tag{5.69}$$

$$\lim_{s \to 0} F(s) = 1 \tag{5.70}$$

$$\lim_{s \to 0} P(s) \neq 0 \; \lim_{s \to 0} |P_d(s)/P(s)| < \infty \tag{5.71}$$

The simplest case that satisfies the condition (5.69) is that $K(s)$ includes an integrator term and $K_f(s)$ does not. Condition (5.70) requires that the sensor is accurate in the steady state. The condition (5.71) requires that the plant is not differentiating and the disturbance is not integrated more times than the manipulated variable. From industrial viewpoint these conditions can be regarded as necessary in order to make the steady-state errors zero robustly [7].

Next, considering a 2DOF PID controller $K(s)$ and $K_f(s)$ satisfying condition (5.69) can be chosen as:

$$K(s) = K_P + \frac{k_I}{s} + k_D s = K_P \left(1 + \frac{1}{sT_{int}} + sT_{der}\right) \tag{5.72}$$

$$K_f(s) = -k_{Pf} - k_{Df} s = -K_P(\alpha + s\beta T_{der}) \tag{5.73}$$

The 2DOF PID controller has five adjustable parameters: k_P, T_{int}, T_{der}, α and β.

To further simplify the treatment we introduce two new assumptions, that are appropriate for many practical problems.

Assumption 1: The sensor has sufficient accuracy and speed as well as the noise can be neglected.

$$F(s) = 1 \; n = 0 \tag{5.74}$$

Assumption 2: The main disturbance enters at the manipulating point.

$$P_d(s) = P(s) \tag{5.75}$$

Under these assumptions (5.70) and (5.71) are satisfied for non-differentiating plants. Since (5.72) and (5.73) satisfy (5.69) when T_{int} is finite, the 2DOF PID controller makes the steady-state errors to a step reference and step disturbance robustly 0 if it is used in the PID or PI configuration.

Fig. 5.20 shows a 2DOF PID control system under Assumptions 1 and 2, which can be transformed equivalently as shown in Fig. 5.21 and Fig. 5.22.

They shall be referred as follows:

- feedforward type Fig. 5.20;
- feedback type Fig. 5.21, where $K_1(s) = K(s) + K_f(s)$ and $K_2(s) = -K_f(s)$;
- set-point filter type Fig. 5.22, where $K_{pf}(s) = K(s) + K_f(s)$.

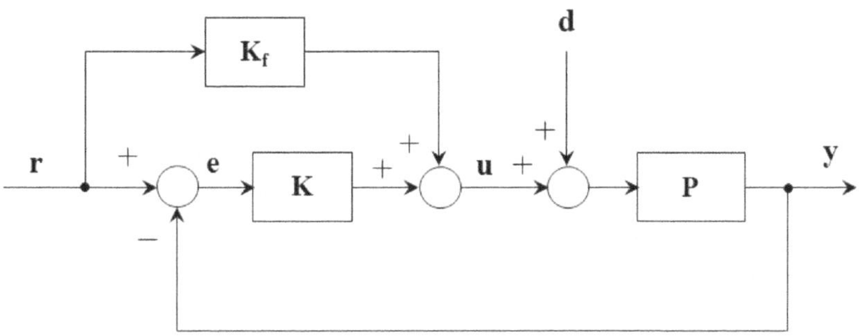

Fig. 5.20 Feedforward type 2DOF control system.

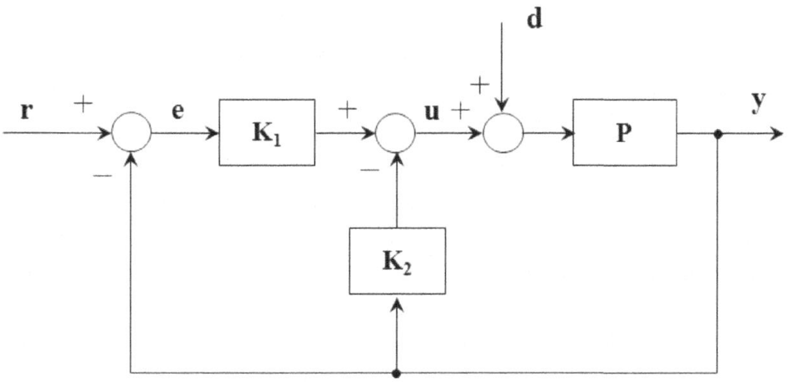

Fig. 5.21 Feedback type 2DOF control system.

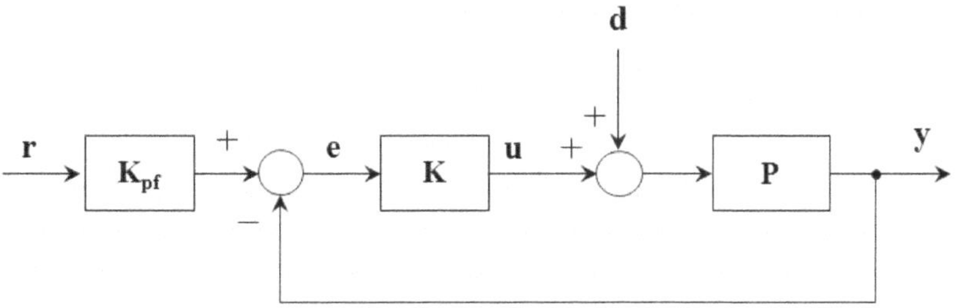

Fig. 5.22 Set-point filter type 2DOF control system.

The Advantages of 2DOF Structures

The response of the controlled variable y to the unit change of the set-point variable r and to the unit step disturbance d are called 'set-point response' and 'disturbance response'. They have been traditionally used as measures of the performance in tuning the PID controllers.

To illustrate the effects of the 2DOF structure let us consider a conventional 1DOF structure under Assumptions 1 and 2.

The closed-loop transfer function of this control system from the set-point variable r to the controlled variable y as well as from disturbance d to y are:

$$G_{YR1} = \frac{PK}{1+PK} \tag{5.76}$$

$$G_{YD1} = \frac{P}{1+PK} \tag{5.77}$$

Here, the subscript $'1'$ means that the quantities are of the 1DOF control system. These two transfer functions include only one tunable element $K(s)$ so they cannot be changed independently.

To be more specific, the two functions are bounded by:

$$G_{YR1}P + G_{YD1} = P \tag{5.78}$$

This equation shows explicitly that for a given $P(s)$, G_{YR1} is uniquely determined if G_{YD1} is chosen, and vice versa. That means if the disturbance response is optimised, the set-point response is often found to be unsatisfactory, and vice versa. This situation is illustrated by a numerical example.

Example 5.5. *Let us consider the following plant:*

$$P(s) = \frac{1}{s+1}e^{-0.2s} \tag{5.79}$$

Considering a 1DOF PID structure, the parameters of the controller can be obtained by the Chien-Hrones-Reswick formula, see previous chapter.

The step response of the first-order plant with time delay leads to:

$$a = 0.19 \; T = 0.96 \; T_d = 0.2 \tag{5.80}$$

The parameters of PID controller are tuned for the quickest disturbance rejection with 20% overshoot:

$$K_P = 6.32 \; T_{int} = 0.4 \; T_{der} = 0.084 \tag{5.81}$$

For the above parameter setting, the closed-loop responses is given by the solid line (see Fig. 5.23). The disturbance response is optimal, however the set-point response has overshoot larger than 50%.

On the other hand, the parameters for the quickest set-point regulation with 20% overshoot are:

$$K_P = 5 \; T_{int} = 1.35 \; T_{der} = 0.094 \tag{5.82}$$

For this parameter setting, the closed-loop responses are represented by the dotted line (see Fig. 5.23). Now, the set-point response is fine, but the disturbance response deteriorates substantially.

Therefore, we cannot optimize the set-point response and the disturbance response at once. However, considering a 2DOF PID control structure will enable us to make both set-point response and disturbance response practically optimal at once. This is illustrated as follows.

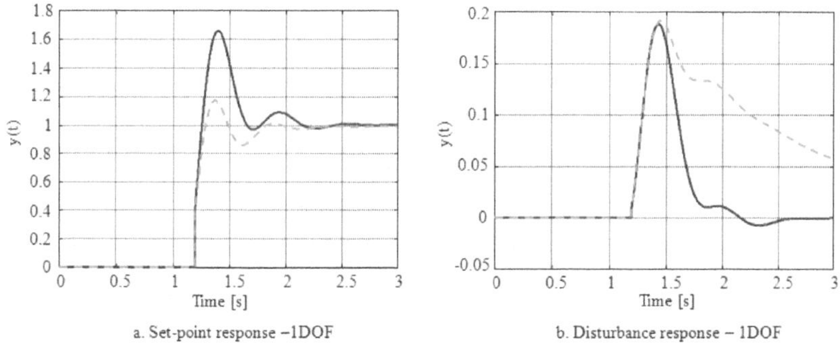

Fig. 5.23 Response of the 1DOF control system.

The closed-loop transfer functions of the 2DOF control system are related to those of the 1DOF control system, by:

$$G_{YR2} = G_{YR1} + \frac{P(s)K_f(s)}{1 + P(s)K(s)} \tag{5.83}$$

$$G_{YD2} = G_{YD1} \tag{5.84}$$

Assuming that $K(s)$ is the same in both control systems, we observe:

- the disturbance responses of the two PID control systems are the same
- the set-point responses differ by the amount of the second term, which can be changed by $K_f(s)$.

Therefore, the set-point response is improved without deteriorating the disturbance response if we use a 2DOF control structure and tune $K_f(s)$.

Let us consider the previous numerical example. The disturbance optimal parameters of the 1DOF system is given by (5.81). If we consider a 2DOF control structure and we choose:

$$\alpha = 0.5 \; and \; \beta = 0.55 \tag{5.85}$$

the set-point response is practically optimal and the disturbance response is unchanged (see Fig. 5.24).

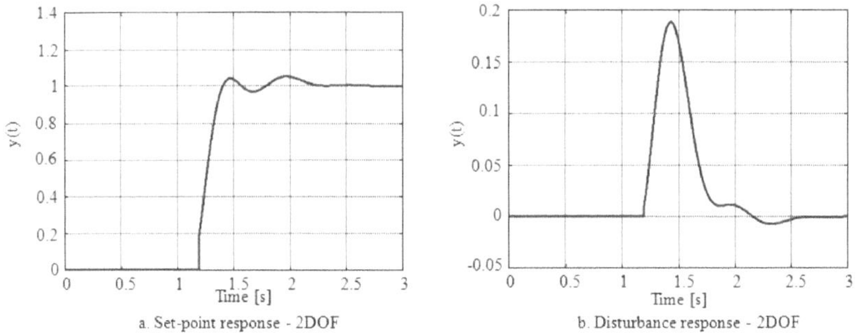

Fig. 5.24 Response of the 2DOF control system.

Therefore, the basic tuning strategy of a 2DOF system can be summarized as:

- STEP 1: optimise the disturbance response by tuning $K(s)$
- STEP 2: let $K(s)$ be fixed and optimise the set-point response by tuning $K_f(s)$, adjusting the parameters α and β.

As conclusion the 2DOF PID controller can optimise simultaneously the disturbance response as well as the set-point response. Its tuning is simple and can be made in two steps.

This problem was not very important in the early days of PID application when the change of the set-point variable was not required very often, but is very important in the modern practice of process control where the change of the set-point variable is frequent.

5.5 Control System Design Verification and Validation

In chapter 3, we have emphasized that no mathematical model can describe a real system, there is always uncertainty, which shall be considered during design.

Until now, in this chapter we have seen different controller design techniques developed for the nominal plant. To verify and validate the controller design is recommended to model the plant as a nominal plant and plant uncertainty (e.g. multiplicative uncertainty) as shown in Fig. 5.25 and simulate the control system.

Our verification checklist should contain at least the following items:

- performance specification (settling time, percent of overshoot, disturbance rejection)
- robustness against parameter variations
- phase and gain margins
- the magnitude of control signal

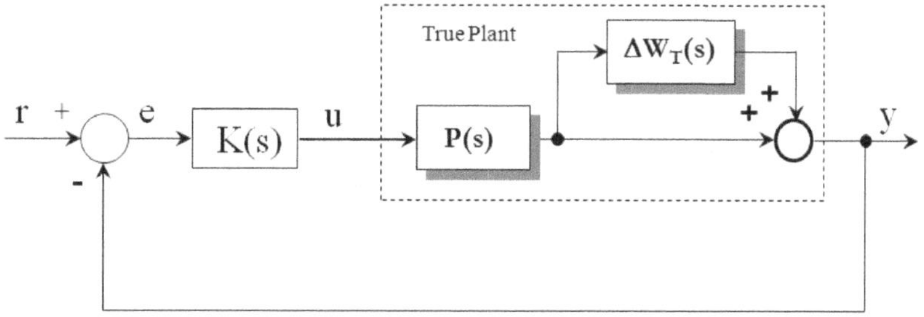

Fig. 5.25 Control system verification using simulation.

- internal signals, when the control loop opens
- internal signals, when the sensor noise is added and increased by a factor of 2.

If all the design requirements are fulfilled the controller design can be validated, but still has to be tested on the real hardware for the real system. We will make remarks about this topic in a later chapter.

Chapter 6

Fundamentals of Robust Control

'A sign of great art is often its openness to multiple readings depending on the observer's perspective.' — Tariq Samad, Former Editor-in-Chief of IEEE Control Systems Magazine

Summary:

This chapter is dedicated to theoretical basics of robust control. In this chapter new concepts are introduced, such as:

- internal stability of the control system
- controller parametrization such that internal stability holds
- robust stability of the control system
- robust performance.

In this chapter the fundamentals of the robust control systems are presented. Until now the controller has been designed for the nominal plant. However, using simulation has been verified and validated considering plant uncertainty, too.

One of the main differences between the classical control design and robust control design is that in case of later one, plant uncertainty is considered explicitly during design. We will try to illustrate the main concepts as well as how the robust control problem is formulated. However, for controller design (for solving the robust control problem) we will need advanced software tools.

In this chapter we will specify the performance of a control system (amplitude of certain signals of interest) using norms. Therefore, first we briefly recall the properties of norms and then we will introduce the $\|\cdot\|_\infty$ norm.

6.1 Review of Norms for Signals and Systems

The properties of a norm are [25]:

- $\|u\| \geq 0$
- $\|u\| = 0 \Leftrightarrow u(t) = 0, \forall t$

- $\|au\| = | a | \|u\|, \forall a \in \Re$
- $\|u + v\| \leq \|u\| + \|v\|$ triangle inequality

6.1.1 Norms for Signals

The **1-norm** of a signal $u(t)$ is the integral of its absolute value:

$$\|u\|_1 := \int_{-\infty}^{\infty} |u(t)| dt \tag{6.1}$$

The **2-norm** of a signal $u(t)$ is:

$$\|u\|_2 := \left(\int_{-\infty}^{\infty} u^2(t) dt \right)^{1/2} \tag{6.2}$$

We remark that the instantaneous power of a signal $u(t)$ is defined $u^2(t)$ and its energy is defined as the square of its 2-norm namely, $\|u\|_2^2$.

The **∞-norm** of a signal is the least upper bound of its absolute value:

$$\|u\|_\infty := \sup_t | u(t) | \tag{6.3}$$

For example, the ∞-norm of the $u(t) = 1 - e^{-t}$ signal equals 1.

6.1.2 Norms for Systems

We consider systems that are linear, time-invariant, casual and finite-dimensional. The system's transfer function is noted by $G(s)$. We introduce two norms for the transfer function $G(s)$.

The **2-norm** of $G(s)$ transfer function is:

$$\|G\|_2 := \left(\frac{1}{2\pi} \int_{-\infty}^{\infty} | G(j\omega) |^2 d\omega \right)^{1/2} \tag{6.4}$$

The **∞-norm** of $G(s)$ transfer function is:

$$\|G\|_\infty := sup | G(j\omega) | \tag{6.5}$$

Theorem 6.1. *The 2-norm of $G(s)$ is finite iff $G(s)$ is strictly proper and has no poles on the imaginary axis; the ∞-norm is finite iff $G(s)$ is proper and has no poles on the imaginary axis.*

How to compute the ∞-norm ?

This requires a search. Set up a fine grid of frequency points $\omega_1, ..., \omega_N$. Then an estimate $\|G(s)\|_\infty$, which is $max | G(j\omega) |$. Another way to find the maximum of $| G(j\omega) |$ is solving the equation:

$$\frac{d | G(s) |^2}{d\omega}(j\omega) = 0 \tag{6.6}$$

Example 6.1

$$G(s) = (as+1)/(bs+1);\ with\ a, b > 0 \tag{6.7}$$

Then

$$\|G(s)\|_\infty = \begin{cases} a/b, & a \geq b \\ 1, & a < b \end{cases} \tag{6.8}$$

The infinity norm in the case of $G(s)$ transfer function is illustrated in Fig. 6.1.

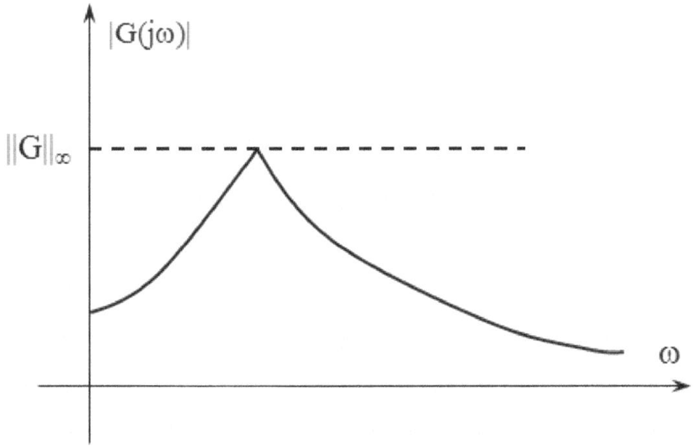

Fig. 6.1 The infinity norm.

In the framework of Matlab's Robust Control Toolbox, this can be done using the commands:

`normh2(a,b,c,d)` - computes the 2-norm for a system with a state-space representation;

`normhinf(a,b,c,d)` - computes the ∞-norm for a system with a state-space representation.

Remark 6.1. *The magnitude of the frequency response $G(j\omega)$ is described as the ratio between the output y and the input u. This seems to be a reasonable way to measure the gain. The problem is, however, that any norm $\|G(j\omega)u(j\omega)\|$ depends not only of the magnitude of the input $\|u\|$ but also on the direction of the input vector. This fact is one of the fundamental differences between single input single output (SISO) and multi input multi output (MIMO) systems, since the input for SISO system is always scalar. Therefore, for MIMO systems the analysis of $G(s)$ transfer function can be made more specific in terms of singular value decomposition of $G(s)$.*

In the next section we are going to introduce the concept of internal stability, which later on is extended to robust stability.

6.2 Internal Stability

Stability is a key issue in control system design. A control system, which provides high-performance in nominal operating conditions and becomes unstable when the system parameters undergo variations has no practical value [25].

The basic feedback loop has three components: a plant (the object to be controlled), a sensor to measure the output of the plant, and a controller to generate the plant's input (usually, actuators are lumped in with the plant).

Let us consider the control system presented in Fig. 6.2 with usual notations:

r - reference or command input
v - sensor output
u - actuating signal, plant input
d - external disturbance
y - plant output
n - sensor noise
i_1, i_2 and i_3 - internal signals.

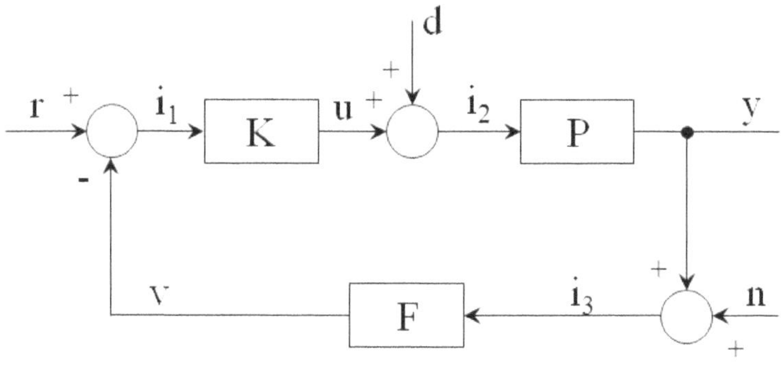

Fig. 6.2 The basic feedback loop.

Next, using the transfer function representation we express the internal signals:

$$\begin{aligned} i_1 &= r - Fi_3 \\ i_2 &= d + Ki_1 \\ i_3 &= n + Pi_2 \end{aligned} \tag{6.9}$$

as well in matrix form:

$$\begin{bmatrix} 1 & 0 & F \\ -K & 1 & 0 \\ 0 & -P & 1 \end{bmatrix} \begin{bmatrix} i_1 \\ i_2 \\ i_3 \end{bmatrix} = \begin{bmatrix} r \\ d \\ n \end{bmatrix} \tag{6.10}$$

Definition 6.1. *The system is* **well-posed** *iff (if and only if) the previous 3x3 matrix is nonsingular, that is the determinant* $1+PKF$ *is not identically equal to zero.*

$$\begin{bmatrix} i_1 \\ i_2 \\ i_3 \end{bmatrix} = \frac{1}{1+PKF} \begin{bmatrix} 1 & -PF & -F \\ K & 1 & -KF \\ PK & P & 1 \end{bmatrix} \begin{bmatrix} r \\ d \\ n \end{bmatrix} \tag{6.11}$$

When P, K and F are proper, a stronger notion of well-posedness is that the nine transfer function above are proper.

Definition 6.2. *The feedback system is automatically* **well-posed, in the stronger sense,** *if* P, K *and* F *are proper and one is strictly proper.*

Definition 6.3. *If the nine transfer functions in the* 3*x*3 *matrix are stable then the feedback system is said to be* **internally stable**. *So internal stability guarantees bounded internal signals for all bounded exogenous signals.*

The idea behind this definition of internal stability is that it is not enough to look only at the input-output transfer functions, such as from r to y.

This transfer function could be stable, so that y is bounded when r is, and yet an internal signal could be unbounded, probably causing internal damage to the physical system.

Control engineers are reluctant to use unstable controllers, especially if the plant itself is stable. System integrity is the motivation. For example, even if a sensor or actuator fails, or is deliberately turned off during start-up or shutdown, thus the feedback loop opens, overall stability should be maintained if both plant and controller individually are stable [25].

Example 6.2

$$K(s) = \frac{s-1}{s+1}; P(s) = \frac{1}{(s^2-1)(s+2)}; F(s) = 1 \tag{6.12}$$

The transfer function from r *to* y *is stable, but from* d *to* y *is not. This is caused by the cancellation of the controller zero and the plant pole at the point* $s=1$.

Theorem 6.2. *The feedback system is internally stable iff the following two conditions hold:*

- *The transfer function* $1+PKF$ *has no zeros in the Res* ≥ 0
- *There is no pole-zero cancellation in Res* ≥ 0 *when the product PKF is formed.*

Furthermore, we introduce the small gain theorem, which plays an important role in robust control theory.

Theorem 6.3. *(Small gain theorem) Suppose that P, $K \in RH_\infty$, then the system in Fig. 6.3 is internally stable, that is*

$$\begin{bmatrix} K(1+PK)^{-1} & (1+PK)^{-1} \\ (1+PK)^{-1} & -P(1+PK)^{-1} \end{bmatrix} \in RH_\infty \qquad (6.13)$$

if $\|K\|_\infty \|P\|_\infty < 1$. In the equation above the input signals are denoted by u_1, u_2 and the output signals by y_1, y_2.

The idea behind is to decompose the system such that it appears to be an interconnection of two subsystems through input/output connection, then the small gain theorem can be applied to analyze its stability.

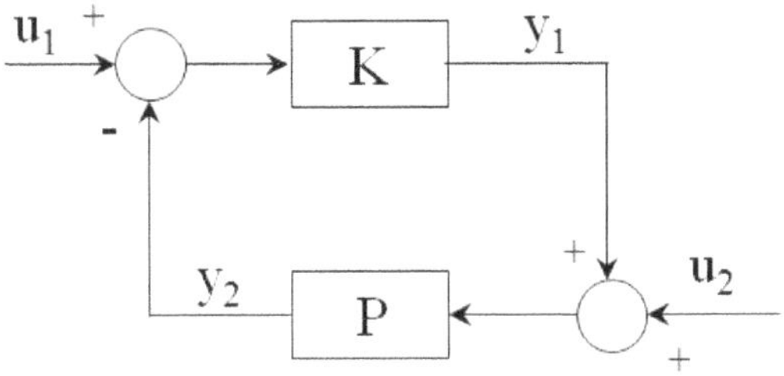

Fig. 6.3 Feedback system.

Next, we will review the Nyquist criterion to check the closed-loop stability based on the inspection of the loop gain $L(s) = K(s)P(s)$, without computing the closed-loop poles (i.e. the roots of $1 + L(s) = 0$).

The Nyquist criterion says: the closed-loop system with loop gain $L(s)$ and negative feedback is stable if and only if the complete Nyquist plot of $L(j\omega)$ encircles the critical point $s = -1$ exactly N_P anticlockwise times in the complex plane, where N_P is the number of unstable poles of $L(s)$.

We will see later in this chapter that robust control does not use the classical gain and phase margins. The main reasons are:

- there are no analytical techniques to optimize the classical gain and phase margin
- there are situations when the classical margins indicate a good robustness against individual gain and phase tolerances, however the feedback loop is not at all robust against simultaneous variations of phase and gain.

The preferred margin in robust control is the so-called critical distance d_{crit} between the Nyquist plot of $L(j\omega)$ and the critical point $s = -1$.

$$d_{crit} = min|L(j\omega) - (-1)| = min|1 + L(j\omega)| \ \forall\omega \qquad (6.14)$$

From the equation above results that the critical distance is the reciprocal of the sensitivity function peak:

$$d_{crit} = \frac{1}{\|S(s)\|_\infty} \tag{6.15}$$

since $S(j\omega) = 1/(1 + L(j\omega)$.

That means, maximizing the critical distance d_{crit} - in terms of robust control - corresponds to minimizing the infinity norm of the sensitivity function $\|S(s)\|_\infty$.

6.2.1 Youla Parametrization

Let us consider the control system with unity feedback shown in Fig. 6.4, where P is strictly proper and K is proper.

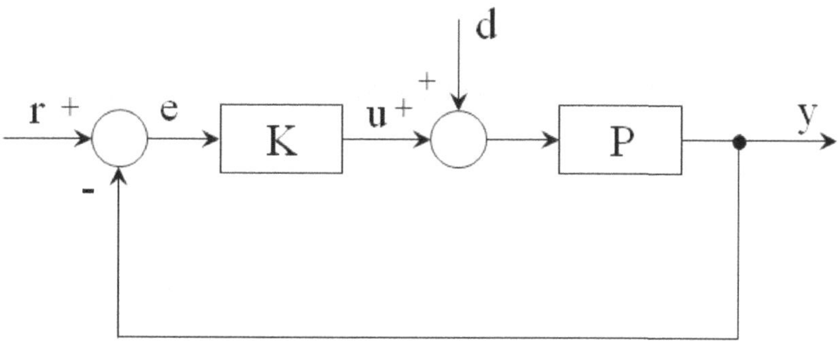

Fig. 6.4 Unity feedback loop.

Taking into account the newly introduced internal stability concept we can formulate the control problem as: **Given** P**, design** K **so that the feedback system is internally stable and for example the output** y **asymptotically tracks a step input** r.

The basic idea is to parametrize all controllers for which internal stability holds (this is called Youla parametrization), and then see if there exists a parameter for which the performance specifications hold.

Definition 6.4. *By definition a plant is* **strongly stabilizable** *if internal stabilization can be achieved with a controller, which is stable.*

Theorem 6.4. P *is strongly stabilizable iff it has an even number of real poles between every pair of real zeros in* $Re s \geq 0$.

Example 6.3

$$P(s) = \frac{(s-1)}{(s+1)(s-2)(s-3)} \tag{6.16}$$

is strongly stabilizable.

One the positive real axis, including ∞, $P(s)$ has two zeros, one at $s = 1$ and one at $s = \infty$. We remark that we have to count only the real and distinct zeros in $Res \geq 0$. Between zeros at $s = 1$ and ∞ lie two poles, at $s = 2$ and $s = 3$. So $P(s)$ is strongly stabilizable.

Controller Parametrization: Stable Plant

We introduce the symbol $\mathcal{S}$ for the family of all stable, proper, real-rational functions.

Theorem 6.5. *Assume that $P \in \mathcal{S}$. The set of all controllers K for which the feedback system is internally stable equals:*

$$K \in \left\{ \frac{Q}{1 - PQ} : Q \in \mathcal{S} \right\} \tag{6.17}$$

where Q is the transfer function from r to u.

Controller Parametrization: Non-stable Plant

P is not stable, let us write P as the ratio of coprime polynomials.

$$P = \frac{N}{D} \tag{6.18}$$

By Euclid's algorithm there are two other polynomials X, Y satisfying the equation

$$NX + DY = 1 \tag{6.19}$$

then is possible to prove the following theorem [25].

Theorem 6.6. *The set of all controllers for which the feedback system is internally stable equals:*

$$K \in \left\{ \frac{X + DQ}{Y - NQ} : Q \in \mathcal{S} \right\} \tag{6.20}$$

To keep things simple, let us present an example of Youla parametrization for a stable plant.

Example 6.4. *[25] Find an internally stabilizing controller for the plant:*

$$P = \frac{1}{(s+1)(s+2)} \tag{6.21}$$

so that y asymptotically tracks a ramp.

All internally stabilizing controllers are:

$$K = \frac{Q}{1 - PQ} \tag{6.22}$$

Since we would like to asymptotically track a ramp the sensitivity transfer function $S(s)$ (from r to e) must have at least two zeros (so Q should have 2 parameters). Let us consider:

$$Q(s) = \frac{as + b}{s + 1} \tag{6.23}$$

This belongs to $\mathcal{S}$ (stable, proper and real-rational functions) and has two variables, a and b, for the assignment of the two zeros. We have:

$$S(s) = 1 - T(s) = 1 - P(s)Q(s) = 1 - \frac{as + b}{(s+1)^2(s+2)} = \frac{s^3 + 4s^2 + (5-a)s + (2-b)}{(s+1)^2(s+2)} \tag{6.24}$$

so we should take $a = 5$, $b = 2$. This gives:

$$Q(s) = \frac{5s + 2}{s + 1} \tag{6.25}$$

Therefore, the controller is:

$$K(s) = \frac{(5s+2)(s+1)(s+2)}{s^2(s+4)} \tag{6.26}$$

The controller is internally stabilizing and has two poles at $s = 0$.

At this moment we are one step closer to our main objective, the designed controller will assure internal stability as well the control system will satisfy the performance specifications. However, at this stage no details are given how the $Q(s)$ transfer function is constructed.

6.2.2 Youla Parametrization for Systems with Time Delays

In this section we will see the connection between the Youla parametrization and the Smith predictor dealing with stable plants with time delay in the form:

$$P(s) = P_0(s)e^{-sT_d} \tag{6.27}$$

where the time delay is $T_d > 0$ and $P_0(s)$ is the nominal plant model for $T_d = 0$.

To obtain a stabilizing controller for $P(s)$, first design a controller $K_0(s)$ that stabilizes the non-delayed feedback system $P_0(s)$. According to the Youla parametrization the $Q_0(s)$ transfer function, has the form:

$$Q_0(s) = \frac{K_0}{1 + P_0 K_0} \tag{6.28}$$

If we use $Q(s) = Q_0(s)$ in the parametrization of stabilizing controllers for $P(s)$ we obtain:

$$K(s) = \frac{Q_0(s)}{1 - P(s)Q_0(s)} = \frac{Q_0(s)}{1 - e^{-sT_d}P_0(s)Q_o(s)} = \frac{K_0(s)}{1 + K_0P_0(1 - e^{-sT_d})} \tag{6.29}$$

The controller structure is exactly the Smith predictor. As been mentioned earlier, the advantage of this approach is that $K_0(s)$ is designed independently of time delay.

Therefore, we can conclude that robust control synthesis methods are powerful tools and can be applied for systems with time delays, too.

Until now, the control design has been performed only for the nominal plant. In the next section we will extend the plant P such that belongs to a set.

In chapter three we have derived using system identification techniques a nominal plant and a plant uncertainty. Depending on the uncertainty model, which we considered the uncertainty has been bound by:

- $W_T(s)$ transfer function, in case of multiplicative uncertainty model
- $W_Q(s)$ transfer function, in case of additive uncertainty model.

We will see in the next section, how internal stability is extended to robust stability considering the plant uncertainty.

6.3 Unstructured Plant Uncertainties

In the chapter dedicated to system identification and model-order reduction we have seen that there is always plant uncertainty. The system identification experiment is valid in a limited frequency range and the derived model is a simplified representation of the real plant. We might divide the plant uncertainties in two categories:

- structured parameter uncertainties due to variation of the physical parameters of the plant. Such parameter variations are induced by: temperature variations, manufacturing dispersion, components aging, etc.
- unstructured parameter uncertainties which are due to neglected high-frequency dynamics. If unstructured parameter uncertainty is not considered during design, we might face the so-called spillover effect. The controller designed neglecting the high-frequency dynamics might not fulfill the specifications when is tested for the real plant.

In the next sections we will briefly recall frequently used plant uncertainty models. We will mainly focus on the unstructured uncertainty models. In some cases structured plant uncertainty can be easily transformed into unstructured plant uncertainty however might make the design process very conservative. Structured plant uncertainty can be handled in a better way in the framework of μ-synthesis, however this exceeds the established framework of the book, so it is skipped.

In case of unstructured plant uncertainty we distinguished between additive and multiplicative plant uncertainty.

6.3.1 Unstructured Additive Plant Uncertainty

The additive plant uncertainty model can be represented in two different forms as described below:

Additive and additive inverse uncertainty:

$$\begin{aligned} \tilde{P} &= P + \Delta W_Q \\ \tilde{P} &= P/(1 + \Delta W_Q P) \end{aligned} \tag{6.30}$$

as shown in Fig. 6.5.

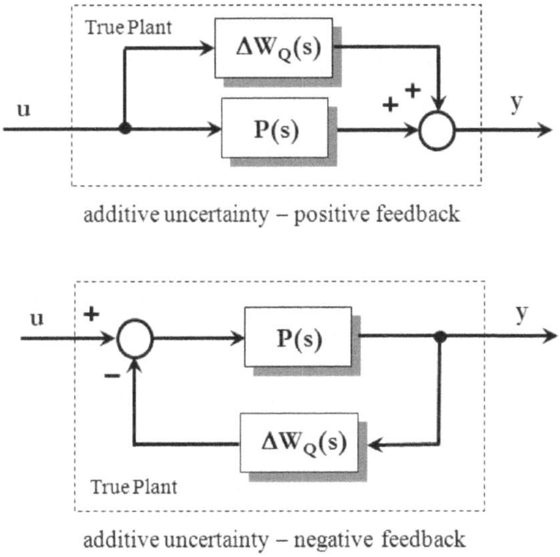

Fig. 6.5 Additive plant uncertainty model.

6.3.2 Unstructured Multiplicative Plant Uncertainty

The multiplicative plant uncertainty can be input or output multiplicative uncertainty

$$\begin{aligned} \tilde{P} &= (1 + \Delta W_T)P \\ \tilde{P} &= P(1 + \Delta W_T) \end{aligned} \tag{6.31}$$

as shown in Fig. 6.6.

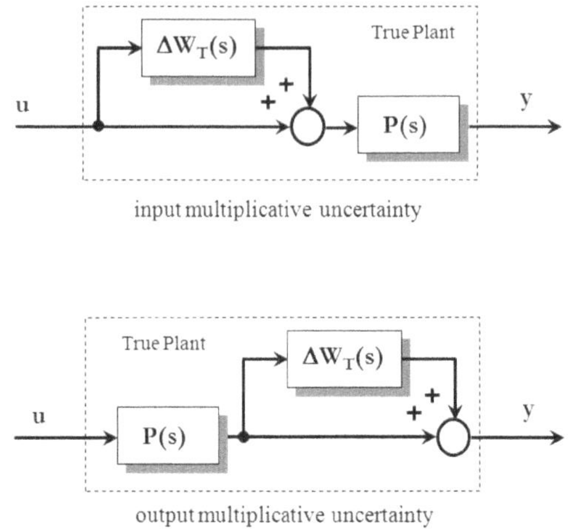

Fig. 6.6 Multiplicative plant uncertainty model.

We might have also input or output inverse multiplicative uncertainty:

$$\tilde{P} = P/(1 + \Delta W_T) \tag{6.32}$$

as shown in Fig. 6.7.

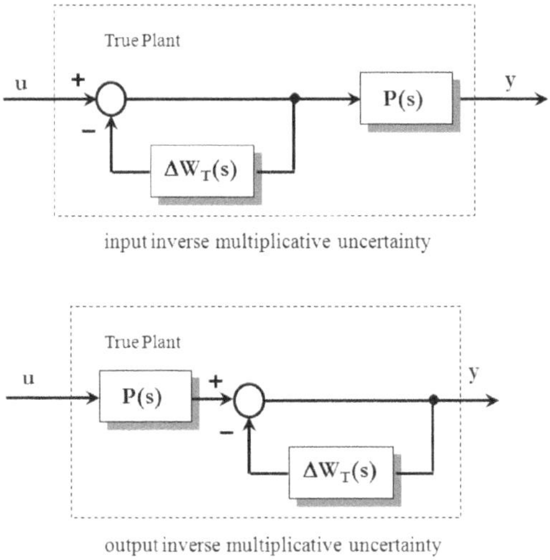

Fig. 6.7 Multiplicative inverse plant uncertainty model.

6.4 Robust Stability

Suppose that the plant transfer function P belongs to a set $\mathcal{P}$, as in the previous section.

Definition 6.5. *A controller K provides* **robust stability** *if it provides internal stability for every plant in $\mathcal{P}$.*

Example 6.5. Stability margin from a robust viewpoint

Consider a multiplicative plant uncertainty:

$$\tilde{P} = (1 + \Delta W_T)P \tag{6.33}$$

Fix a positive number β and consider the family of plants:

$$\{\tilde{P} : \Delta W_T \textit{ stable and } \|\Delta\|_\infty \leq \beta\} \tag{6.34}$$

Now a controller K that achieves internal stability for the nominal plant P will stabilize this entire family if β is small enough. Denote by β_{sup} the least upper bound on β such that K achieves internal stability for the entire family. Then β_{sup} is a stability margin (with respect to this uncertainty model) [25].

Classical Measures of Stability Margin

Gain margin:
Perturb the plant to kP, where k is a positive real number. The upper gain margin, denoted k_{max}, is the first value of k greater than 1 when the feedback system is not internally stable. Similarly, the lower gain margin, k_{min}, is the least non-negative number such that internal stability holds for $k_{min} \leq k \leq k_{max}$.

Phase margin:
Now perturb the plant to $e^{j\phi}P$ with ϕ, a positive real number. The phase margin, ϕ_{max}, is the maximum number (usually expressed in degrees) such that internal stability holds for $0 < \phi \leq \phi_{max}$

It is generally found that gain margins of 10-20 dB or more combined with phase margins between 40 and 60 degrees result in reasonable tradeoffs between bandwidth and stability.

Remark 6.2. *Gain and phase margins have traditionally been important measures of stability. Notice, however, that the gain and phase margins can be relatively large but simultaneous small changes in gain and phase could cause internal instability.*

6.4.1 Robust Stability for Different Uncertainty Models

Theorem 6.7. *Considering an additive uncertainty model bounded as* $|\Delta_a(j\omega)| \leq |W_Q(j\omega)|$ *the controller* K *provides* **robust stability** *iff* $\|W_Q KS\|_\infty < 1$.

Proof - additive uncertainty bound:
Given a nominal plant $P(s)$ and an additive uncertainty bound $|\Delta_a(s)| \leq |W_Q(s)|$ characterizing the set of all possible plants $\tilde{P}(s)$, we want to derive the condition under which the controller $K(s)$ assures robust stability. We assume that $P(s)$ and $\tilde{P}(s)$ have the same number of poles in the right-half complex plane.

$$\tilde{P}(s) = P(s) + \Delta_a(s) \tag{6.35}$$

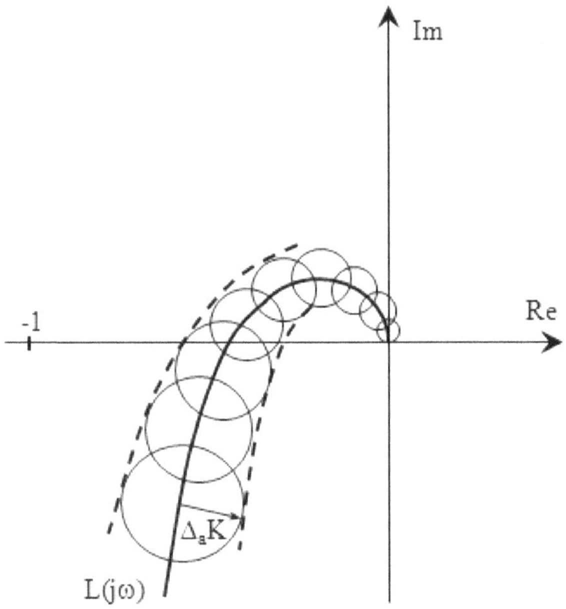

Fig. 6.8 Robust stability test using the Nyquist plot.

Let's assume that the controller $K(s)$ assures internal stability for the nominal plant $P(s)$. The Nyquist curve $L(j\omega) = P(j\omega)K(j\omega)$ is presented in Fig. 6.8.

The Nyquist plot for $\tilde{P}$ lie within a tube around the nominal Nyquist curve.

$$\tilde{L}(j\omega) = L(j\omega) + \Delta_a(j\omega)K(j\omega) \tag{6.36}$$

The radius of the tube is changing with the frequency, as shown in Fig. 6.8. Then, we can conclude that the system is robustly stable iff the point (-1,0) in the complex plane is outside of this tube, which means:

$$|\tilde{L}(j\omega) - (-1)| > 0 \; \forall \omega \; and \; \forall \Delta_a(j\omega) \tag{6.37}$$

Then we can write:

$$|L(j\omega) + \Delta_a(j\omega)K(j\omega) + 1| > 0 \tag{6.38}$$

It is sufficient to prove that:

$$|1 + P(j\omega)K(j\omega)| - |\Delta_a(j\omega)K(j\omega)| > 0 \tag{6.39}$$

Using the additive uncertainty bound $|\Delta_a(j\omega)| \leq |W_Q(j\omega)|$ we can write:

$$|1 + P(j\omega)K(j\omega)| \geq |W_Q(j\omega)K(j\omega)| \tag{6.40}$$

The previous inequality can be written in a compact form as:

$$|W_Q K(1 + PK)^{-1}| < 1 \tag{6.41}$$

or using the infinity norm and the sensitivity transfer function, it becomes:

$$\|W_Q KS\|_\infty < 1 \tag{6.42}$$

Theorem 6.8. *Considering a multiplicative uncertainty model bounded as* $|\Delta_m(j\omega)| \leq |W_T(j\omega)|$ *the controller* K *provides* **robust stability** *iff* $\|W_T T\|_\infty < 1$.

Proof - multiplicative uncertainty bound:
If a multiplicative uncertainty bound $W_T(j\omega)$ is given instead of the additive uncertainty bound:

$$|\Delta_m(j\omega)| < |W_T(j\omega)| \; \forall \omega \tag{6.43}$$

then

$$\tilde{P} = P(1 + \Delta_m) = P + \Delta_a = P(1 + P^{-1}\Delta_a) \tag{6.44}$$

Furthermore,

$$|\Delta_m(j\omega)| = |P^{-1}(j\omega)\Delta_a(j\omega)| \leq |P^{-1}W_Q(j\omega)| \; \forall \omega \tag{6.45}$$

then the relation between the additive and multiplicative uncertainty bounds is:

$$|\Delta_m(j\omega)| = |P^{-1}(j\omega)W_Q(j\omega)| \; \forall \omega \tag{6.46}$$

From equations (6.41) and (6.46) we obtain:

$$|W_Q K(1 + PK)^{-1}| = |P^{-1}W_Q PK(1 + PK)^{-1}| = |W_T T| < 1 \tag{6.47}$$

or using the infinity norm:

$$\|W_T T\|_\infty < 1 \tag{6.48}$$

Table 6.1 summarizes the robust stability test for other uncertainty models, too.

Table 6.1 Summary of robust stability test.

Uncertainty Model	Robust Stability Condition
$(1+\Delta W_T)P$ - multiplicative uncertainty	$\|W_T T\|_\infty < 1$
$P+\Delta W_Q$ - additive uncertainty	$\|W_Q KS\|_\infty < 1$
$P/(1+\Delta W_T P)$	$\|W_T PS\|_\infty < 1$
$P/(1+\Delta W_Q)$	$\|W_Q S\|_\infty < 1$

Remark 6.3. *When we construct the upper bound of the uncertainty described by $W_Q(s)$ and/or $W_T(s)$ transfer functions only the magnitude of these functions are relevant.*

6.5 Robust Performance

The transfer function from the reference r to tracking error e is $S(s)$, which is the well-known sensitivity function.

In the previous chapters performance specifications have been defined in terms of settling time and percent of overshoot of a second-order system. These specifications have been translated into an ideal complementary sensitivity transfer function $T_{id}(s)$ and implicitly into an ideal sensitivity transfer function: $S_{id}(s) = 1 - T_{id}(s)$.

Next, we will try to express the performance specifications in terms of sensitivity transfer function using the infinity norm.

Let us suppose that the input signal r can be any sinusoid of amplitude ≤ 1 and we want the tracking error e to have amplitude less than ϵ. Then the performance specification can be expressed as:

$$\|S\|_\infty < \epsilon \tag{6.49}$$

The maximum amplitude of e equals the ∞-norm of the transfer function. Or if we define the (trivial, in this case) weighting function $W_S(s) = 1/\epsilon$ then the performance specification is:

$$\|W_S S\|_\infty < 1 \tag{6.50}$$

The situation becomes more realistic and more interesting with a frequency dependent weighting function. As a remark only the magnitude of $W_S(j\omega)$ is relevant, so any poles or zeros can be reflected into the left half-plane without changing the magnitude.

In several applications, for example aircraft flight-control design, designers have acquired through experience, desired shapes for the Bode magnitude plot of $S(s)$.

In particular, suppose that good performance is known to be achieved if the plot $\mid S(j\omega) \mid$ lies under some curve.

We could rewrite this as:

$$\mid S(j\omega) \mid < \mid W_S(j\omega) \mid^{-1}, \forall\omega \tag{6.51}$$

Remark 6.4. *Other performance specifications could be posed by focusing on the response to the other two exogenous inputs, such as* d *disturbance and* n *noise.*

Now let suppose that the plant transfer function belongs to a set $\mathcal{P}$. The general notion of **robust performance** is that internal stability and nominal performance, should hold for all plants in $\mathcal{P}$.

Additive Uncertainty Model

Theorem 6.9. *A necessary and sufficient condition for robust performance is:*

$$|W_S S| + |W_Q K S| < 1 \tag{6.52}$$

or

$$\| |W_S S| + |W_Q K S| \|_\infty < 1 \tag{6.53}$$

The proof is left for the reader, basically it is very similar with the proof in case of multiplicative uncertainty model, presented next.

Multiplicative Uncertainty Model

Theorem 6.10. *A necessary and sufficient condition for robust performance is:*

$$|W_S S| + |W_T T| < 1 \tag{6.54}$$

or

$$\| \mid W_S S \mid + \mid W_T T \mid \|_\infty < 1 \tag{6.55}$$

The previous theorem gives the robust performance test under the following conditions:

- multiplicative uncertainty model: $(1 + \Delta W_T)P$;
- nominal performance condition: $\|W_S S\|_\infty < 1$.

Proof:
Consider again the multiplicative uncertainty model. Besides stability, we are also interested in the performance of the closed-loop system. The nominal performance of the control system can be measured in terms of tracking error and can be expressed in a compact form as:

$$\|W_S S\|_\infty < 1 \tag{6.56}$$

If the previous inequality holds for every plant $\tilde{P} = P(1 + \Delta_m)$, then nominal performance condition is called robust performance condition. Of course, this definition makes sense only if the system is robustly stable. Therefore, we say that a controller achieves robust performance if achieves internal stability (see definition of robust stability) and nominal performance for every plant $P \in \tilde{P}$.

Considering a multiplicative uncertainty model with: $|\Delta_m(j\omega)| \leq |W_T(j\omega)|$, the nominal performance condition is written as:

$$|S^{-1}| > |W_S| \tag{6.57}$$

and for every plant $P \in \tilde{P}$ becomes:

$$|1 + PK(1 + \Delta_m)| = |1 + PK| \left| 1 + \Delta_m \frac{PK}{1 + PK} \right| > |W_S| \tag{6.58}$$

The previous inequality can be written as:

$$|W_S S| < |1 + \Delta_m T| \tag{6.59}$$

The right hand side of this inequality is written as $1 - |\Delta_m T| \leq |1 + \Delta_m T|$, then:

$$|W_S S| < 1 - |\Delta_m T| \tag{6.60}$$

and taking into account the multiplicative uncertainty bound becomes:

$$|W_S S| < 1 - |W_T T| \tag{6.61}$$

which rewritten is exactly the robust performance condition:

$$\| \mid W_S S \mid + \mid W_T T \mid \|_\infty < 1 \tag{6.62}$$

6.5.1 Performance and Robustness Bounds

It is common practice to lump the effects of all plant uncertainty into a single multiplicative uncertainty, so that the control design requirements are written:

$$|S^{-1}(j\omega)| \geq |W_S(j\omega)| \; and \; |T(j\omega)| \leq |W_T^{-1}(j\omega)| \tag{6.63}$$

It is interesting to note that in the upper half of the Fig. 6.9 $|1 + L(j\omega)| >> 1$:

$$|L(j\omega)| \approx \frac{1}{|S(j\omega)|} \tag{6.64}$$

while in the lower half of the Fig. 6.9:

$$|L(j\omega)| \approx |T(j\omega)| \tag{6.65}$$

This results from the fact that:

$$S(s) = 1/(1 + L(s)) \approx 1/L(s) \; if \; |L(j\omega)| >> 1 \tag{6.66}$$

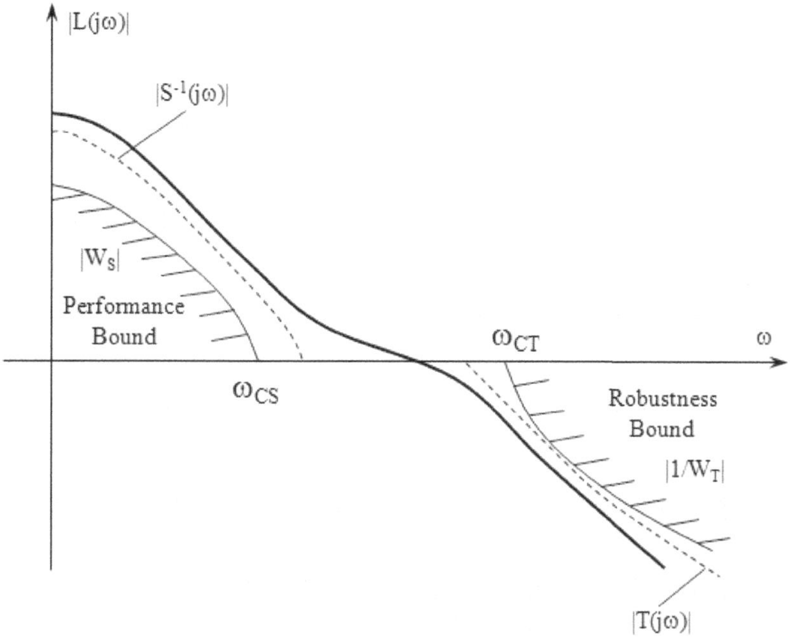

Fig. 6.9 Performance and robustness bounds.

as well as

$$T(s) = L(s)/(1 + L(s)) \approx L(s) \; if \; |L(j\omega)| << 1 \tag{6.67}$$

Thus, it is not uncommon to see specifications on disturbance attenuation and stability margin expressed directly in terms of forbidden regions for the Bode plot.

An important point to note in choosing design specifications $W_S(s)$ and $W_T(s)$ is that the 0 dB crossover frequency of $W_S(s)$ must be sufficiently below the 0 dB crossover frequency of $W_T^{-1}(s)$.

According to the notations used in Fig. 6.10 $\omega_{CS} < \omega_{CT}$, if the previous inequality is not satisfied than robust performance cannot be achieved.

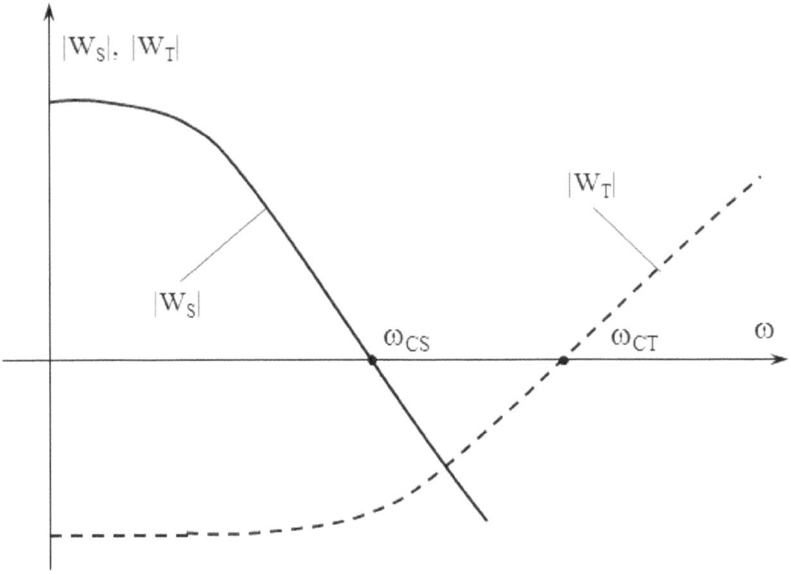

Fig. 6.10 Robustness and performance weighting functions.

6.6 Design for Robust Performance

Sometimes robust stability and nominal performance would be sufficient for certain applications. This is weaker (easier to satisfy) than the robust performance condition. Let us recall that the nominal performance condition is:

$$\|W_S(s)S(s)\|_\infty < 1 \tag{6.68}$$

and the robust stability condition (in the case of multiplicative uncertainty model) is:

$$\|W_T(s)T(s)\|_\infty < 1 \tag{6.69}$$

The robust performance condition is written as:

$$\| \mid W_S S \mid + \mid W_T T \mid \|_\infty < 1 \tag{6.70}$$

Furthermore, we can write:

$$max(|W_S S|, |W_T T|) \leq |W_S S| + |W_T T| \leq 2max(|W_S S|, |W_T T|) \tag{6.71}$$

If nominal performance and robust stability are obtained with a safety factor of 2,

$$\|W_S S\|_\infty < 1/2 \; and \; \|W_T T\|_\infty < 1/2 \tag{6.72}$$

then robust performance is automatically achieved.

A compromise condition (also called modified robust performance condition) is:

$$\| |W_S S|^2 + |W_T T|^2 \|_\infty^{1/2} < 1 \tag{6.73}$$

which is associated with the Euclidean norm, which mathematically is the most tractable.

It is mentioned in [25], that the tradeoffs between $|W_S S|$ and $|W_T T|$ inherent in control problems mean that although the norms may differ by as much as a factor of 2, the actual solutions one gets by using different norms are not very different.

The robust performance condition has a nice graphical interpretation. For each frequency ω we construct two disks: one with center -1, radius $|W_S(j\omega)|$, the other with center $L(j\omega)$, radius $|W_T(j\omega)L(j\omega)|$. Then the robust performance condition holds iff these two disks are disjoint (see Fig. 6.11).

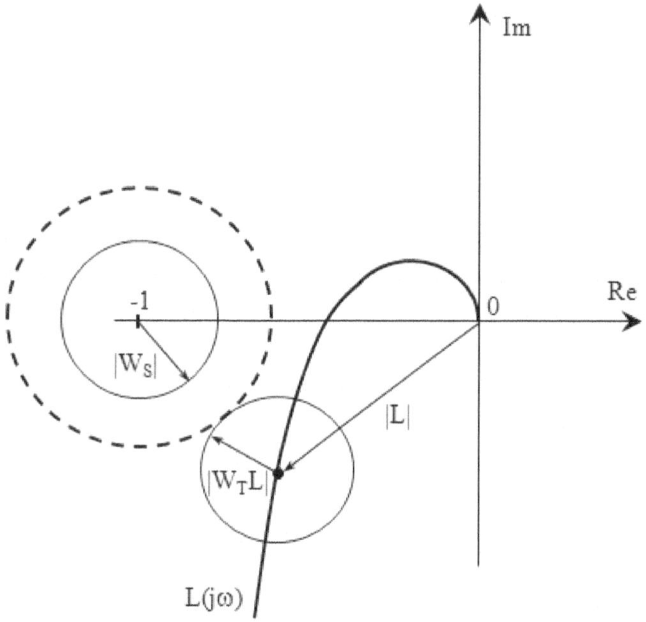

Fig. 6.11 Robust performance condition - graphical interpretation.

Remark 6.5. *In general, the robust performance problem is written, according to equation (6.74):*

$$|W_S(j\omega)S(j\omega)| + |W_T(j\omega)T(j\omega)| < \gamma \ \forall \omega \tag{6.74}$$

where γ is a positive constant and the main objective is minimizing γ over all stabilizing controllers. However, this problem is more difficult than the problem of finding a stabilizing controller and minimizing γ according to equation (6.75):

$$|W_S(j\omega)S(j\omega)|^2 + |W_T(j\omega)T(j\omega)|^2 < \gamma^2 \ \forall\omega \tag{6.75}$$

In the literature, this problem is known as the mixed sensitivity minimization, or two-block $\mathcal{H}_\infty$ control problem.

6.6.1 Performance Weighting Function Design for Tracking Control

In the case of a tracking control problem, a common way to specify desired closed-loop performance is in terms of settling time and percent of overshoot.

According to the classical control theory, in the case of a second-order system, with damping ratio ξ and natural frequency ω_n, the settling time T_{set} and the percent of overshoot $P.O.$ can be expressed approximately as:

$$T_{set} \approx \frac{4}{\xi\omega_n} \tag{6.76}$$

$$P.O. = 100 \cdot e^{-\xi\pi\sqrt{1-\xi^2}} \tag{6.77}$$

We will accomplish the performance specifications by shaping $T(s)$, the transfer function, from r to y, so that it approximates a second-order system. The ideal $T_{id}(s)$ is:

$$T_{id}(s) = \frac{\omega_n^2}{s^2 + 2\xi\omega_n s + \omega_n^2} \tag{6.78}$$

Then the ideal sensitivity function $S_{id}(s)$ is:

$$S_{id}(s) = 1 - T_{id}(s) \tag{6.79}$$

Now it's possible to take the weighting function $W_S(s)$ to be $S_{id}(s)^{-1}$, so that is:

$$W_S(s) = \frac{1}{1 - T_{id}(s)} = \frac{s^2 + 2\xi\omega_n s + \omega_n^2}{s(s + 2\xi\omega_n)} \tag{6.80}$$

One further adjustment is required in the problem setup: $W_S(s)$ must be stable, so the function above is modified to a stable one.

$$W_S(s) = \frac{s^2 + 2\xi\omega_n s + \omega_n^2}{(s + 0.001)(s + 2\xi\omega_n)} \tag{6.81}$$

We remark that the derived $W_S(s)$ is the ideal performance weighting function, thus the robust performance condition cannot be satisfied. However it can represent a good starting point for design iterations.

6.6.2 Performance Weighting Function Design for Disturbance Rejection

In this section we consider a disturbance rejection control problem and the control objective is to make the transfer function from disturbance d to output y denoted by $T_{yd}(s)$ small, in the frequency range where the disturbance d is large.

$$T_{yd}(s) = \frac{P(s)}{1 + P(s)K(s)} = P(s)S(s) \tag{6.82}$$

In other terms minimize the $T_{yd}(s)$ transfer function over a frequency range, according to the infinity norm, which can be expressed as:

$$\|T_{yd}(s)\|_\infty < \rho \tag{6.83}$$

where ρ is a constant, or in a more general case a frequency dependent weighting function $W_P^{-1}(s)$, as follows:

$$|T_{yd}(s)| = |P(s)S(s)| < |W_P(s)|^{-1} \tag{6.84}$$

From a robust control perspective this can be written as:

$$\|W_P(s)P(s)S(s)\|_\infty < 1 \tag{6.85}$$

Since the nominal performance is $\|W_S(s)S(s)\|_\infty < 1$ as a first guess we might choose $W_S(s) = W_P(s)P(s)$.

The simplest approach - in the aim to obtain good disturbance rejection - is to choose the $W_P^{-1}(s)$ weighting function as a constant, in the frequency range where the disturbance is large. Another approach is to shape the $W_P^{-1}(s)$, weighting function as a frequency dependent function.

For example, in case of active vibration suppression of flexible structures the basic goal is to increase the natural damping of the structure. Therefore, the ideal transfer function from input disturbance to control output denoted by $T_{yd}^{id}(s)$ is known in advance.

The ideal sensitivity transfer function $S_{id}(s)$ is:

$$S_{id}(s) = \frac{T_{yd}^{id}(s)}{P_(s)} \tag{6.86}$$

Then a possible performance weighting transfer function is:

$$W_S(s) = \frac{P(s)}{T_{yd}^{id}(s)} \tag{6.87}$$

Example 6.6. *Let us consider a plant with lightly damped poles, given by:*

$$P(s) = \frac{1}{3s^2 + s + 2} \tag{6.88}$$

If our main goal is disturbance rejection then we can choose the performance weighting $W_S(s)$ as:

$$W_S(s) = \frac{3s^2 + 3s + 2}{3s^2 + s + 2} \tag{6.89}$$

6.7 Robust Controller Synthesis Problem

The most general block diagram of a control system is shown in Fig. 6.12.

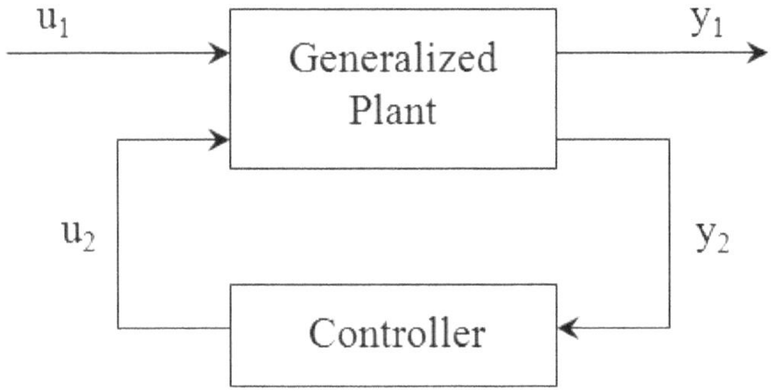

Fig. 6.12 Control problem in a general framework.

- The generalized plant consists of everything that is fixed at the start of the control design exercise: plant, actuators, digital-to-analog converters, sensors, etc.
- The controller consists of the designable part, it may be: electric circuit, programmable logic controller, general-purpose computer, etc.
- The signals u_1 (also noted in the literature with w) are all exogenous inputs: references, disturbances, sensor noises and so on.
- The signals y_1 (also noted in the literature with z) are all the signals we wish to control: tracking errors, actuator signals whose values must be kept between certain limits.
- The vector y_2 provides the input to the feedback controller. Finally, u_2 contains all controlled inputs to the generalized plant.

The **synthesis problem** can now be stated as follows: **Given a set of generalized plants, a set of exogenous inputs, and an upper bound on the size y_1, design an implementable controller to achieve this bound.**

We note that y_1 need not be physically measured, although may be overlap between the signals that comprise y_1 and the measured signals that comprise y_2 so that y_1 may be partly or fully measured.

6.7.1 Mixed-Sensitivity Optimization for Tracking Problem

The classical command-following architecture is shown in Fig. 6.13 with the usual notations, is recast into the general control framework shown in Fig. 6.14.

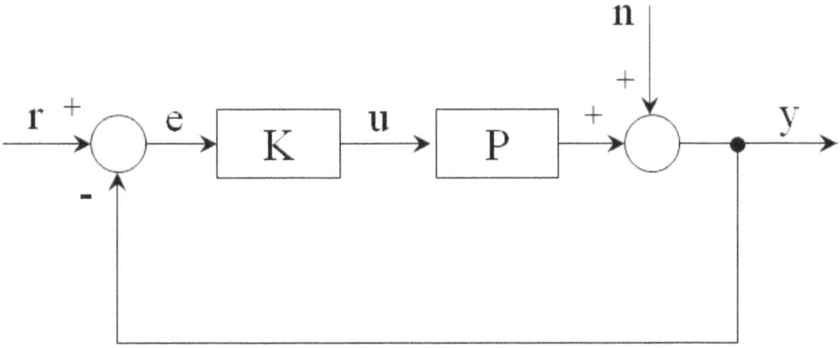

Fig. 6.13 Control structure in classical control framework.

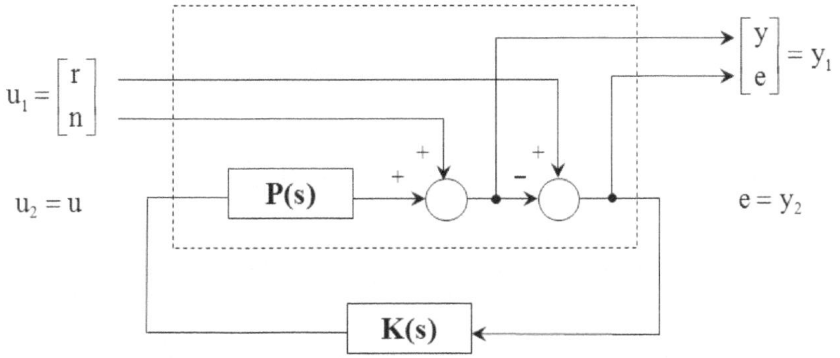

Fig. 6.14 Control structure in general control framework.

Now writing the equations at the summing junctions we have:

$$\begin{aligned} y &= n + Pu \\ e &= r - (n + Pu) \end{aligned} \tag{6.90}$$

which expressed in matrix form:

$$\begin{bmatrix} y \\ e \\ e \end{bmatrix} = \begin{bmatrix} 0 & 1 & P \\ 1 & -1 & -P \\ 1 & -1 & -P \end{bmatrix} \begin{bmatrix} r \\ n \\ u \end{bmatrix} \tag{6.91}$$

the inner structure of the general control architecture is presented in Fig. 6.15

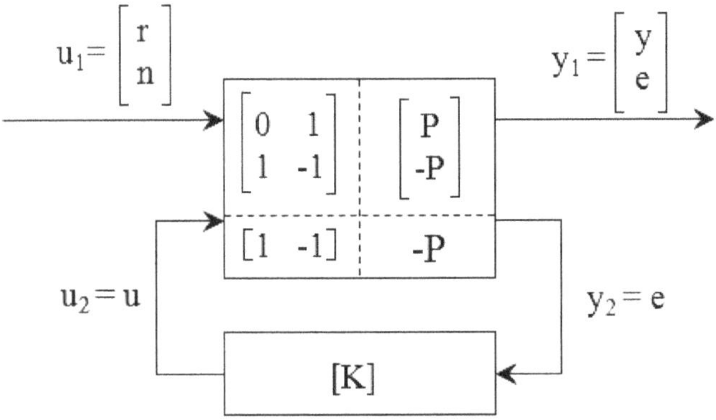

Fig. 6.15 Control architecture - inner structure.

The next step is to express the transfer function between u_1 and y_1, written as $y_1 = T_{y1u1}u_1$. After elementary computation we obtain:

$$\begin{bmatrix} y \\ e \end{bmatrix} = \begin{bmatrix} T(s) & S(s) \\ S(s) & -S(s) \end{bmatrix} \begin{bmatrix} r \\ n \end{bmatrix} \tag{6.92}$$

where $S(s)$ is the sensitivity transfer function and $T(s)$ is the complementary sensitivity transfer function. Therefore T_{y1u1} is written as:

$$T_{y1u1} = \begin{bmatrix} T(s) & S(s) \\ S(s) & -S(s) \end{bmatrix} \tag{6.93}$$

For simplicity we consider the ideal case $n = 0$ then $u_1 = r$ and

$$y_1 = \begin{bmatrix} y \ e \end{bmatrix}^T \tag{6.94}$$

and T_{y1u1} becomes:

$$T_{y1u1} = \begin{bmatrix} T(s) \\ S(s) \end{bmatrix} \tag{6.95}$$

Robust performance specifications can be imposed on the sensitivity $S(s)$, control sensitivity $Q(s)$ and complementary sensitivity $T(s)$ transfer functions as shown in Fig. 6.16.

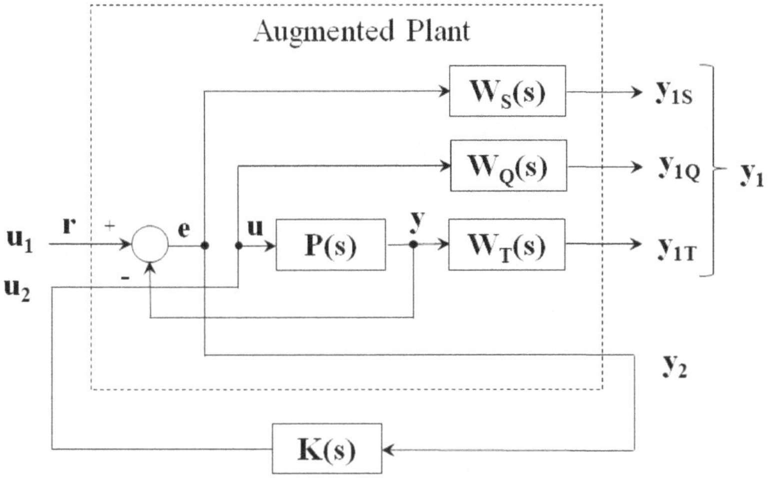

Fig. 6.16 Plant augmentation.

In order to combine all conditions in one equation we augment the plant as shown in Fig. 6.16.

In this way the controller synthesis method has the possibility to consider the pair S, T (or S, Q) with the perspective of deciding at each frequency ω whether S or T (alternatively S or Q) should be made small. Such compensator design algorithms are collectively called mixed sensitivity methods.

The closed-loop transfer function in this approach is:

$$T_{y_1u_1} = \begin{bmatrix} W_S S \\ W_Q K S \\ W_T T \end{bmatrix} \tag{6.96}$$

The transfer functions $P(s), W_S(s)$ and $W_T \cdot P(s)$ (in the framework of Matlab's Robust Control Toolbox) must be proper, i.e. bounded as $s \to \infty$, however W_T may be improper.

The robust controller synthesis problem, considering a mixed sensitivity approach can be stated as:

$$\|T_{y_1u_1}\|_\infty < 1 \tag{6.97}$$

Iterations are performed in order to satisfy the robust performance condition:

$$\|T_{y_1u_1}\|_\infty < \gamma \tag{6.98}$$

with $\gamma < 1$. Practical experience indicates that good designs are obtained with $\gamma \approx 0.5 - 0.8$.

An important notice: in the Matlab's Robust Control Toolbox the role of γ is inverted so that the right-hand side is always normalized to one, so the robust stability condition is written:

$$\|\gamma T_{y_1 u_1}\|_\infty < 1 \tag{6.99}$$

and is satisfied when $\gamma \geq 1$. This might be confusing, therefore requires special attention.

In Matlab's Robust Control Toolbox is possible to scale all the weighting functions (output channels) : $W_S(s)$, $W_Q(s)$ and $W_T(s)$ by γ as follows:

$$W_S(s) = \gamma \cdot W_S(s) \; W_Q(s) = \gamma \cdot W_Q(s) \; W_T(s) = \gamma \cdot W_T(s) \tag{6.100}$$

The default choice in the Matlab's Robust Control Toolbox is to scale all the output channels, usually we will scale only one output channel related to the sensitivity weighting function $W_S(s)$.

The other two weighting functions are related to plant uncertainty - derived using system identification techniques - and are well-defined values in case of a given plant.

During the γ iterations if $\gamma > 1$ means that the robust performance condition is satisfied with the specified performance weighting functions. If $\gamma \leq 1$ means that the robust performance condition is not satisfied, so we have to relax the imposed performance conditions. When all the output channels are scaled the $1/\gamma$ value equals the ∞-norm, otherwise is required to check the ∞-norm using for example the `normhinf` command. These remarks refer to the Matlab's Robust Control Toolbox.

The H_∞ controller design problem can be formulated in different ways, but finally the plant is augmented and its state-space representation is:

$$\begin{aligned} \dot{x} &= Ax + B_1 u_1 + B_2 u_2 \\ y_1 &= C_1 x + D_{11} u_1 + D_{12} u_2 \\ y_2 &= C_2 x + D_{21} u_1 + D_{22} u_2 \end{aligned} \tag{6.101}$$

The augmented plant matrices are usually represented as follows:

$$G(s) = \left[\begin{array}{c|cc} A & B_1 & B_2 \\ \hline C_1 & D_{11} & D_{12} \\ C_2 & D_{21} & D_{22} \end{array}\right] \tag{6.102}$$

A Frequently Encountered Case in Practice

Often in practice, the plant transfer unction $P(s)$ is strictly proper (or can be easily transformed into a strictly proper transfer funcion), however the plant uncertainty described by the $W_T(s)$ transfer function is improper. The Robust Control Toolbox of Matlab requires that the product $W_T(s)P(s)$ to be a proper transfer function.

In this case let us note with:

$$P(s) = \left[\begin{array}{c|c} A_P & B_P \\ \hline C_P & 0 \end{array}\right] \tag{6.103}$$

and with:

$$P(s)W_T(s) = \left[\begin{array}{c|c} A_T & B_T \\ \hline C_T & D_T \end{array}\right] \tag{6.104}$$

It is possible to show (left to the reader, see state-space representation of a transfer function) that $A_T = A_P$ and $B_T = B_P$.

Furthermore, a strictly proper performance weighting transfer function $W_S(s)$ is considered:

$$W_S(s) = \left[\begin{array}{c|c} A_S & B_S \\ \hline C_S & 0 \end{array}\right] \tag{6.105}$$

Next, we form the augmented plant as shown in Fig. 6.17 as well as its state-space representation as shown in Fig. 6.18.

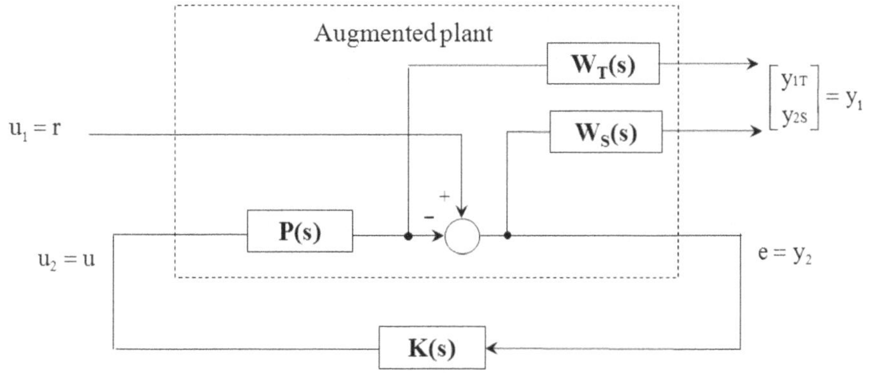

Fig. 6.17 Plant augmentation with $W_S(s)$ and $W_T(s)$.

Then we can write, using state-space representation:

$$\begin{aligned} \dot{x}_P &= A_P x_P + B_p u_2 \\ \dot{x}_S &= A_S x_S + B_S(u_1 - C_P x_P) \\ y_{1S} &= C_S x_S \\ y_{1T} &= C_T x_p + D_T u_2 \\ y_2 &= u_1 - C_P x_p \end{aligned} \tag{6.106}$$

$$\begin{bmatrix} \dot{x}_P \\ \dot{x}_S \\ y_{1S} \\ y_{1T} \\ y_2 \end{bmatrix} = \left[\begin{array}{cc|cc} A_P & 0 & 0 & B_p \\ -B_S C_P & A_S & B_S & 0 \\ \hline 0 & C_S & 0 & 0 \\ C_T & 0 & 0 & D_T \\ -C_P & 0 & I & 0 \end{array}\right] \begin{bmatrix} x_P \\ x_S \\ u_1 \\ u_2 \end{bmatrix} \tag{6.107}$$

Example 6.7. *Consider the augmented plant according to Fig. 6.17, with* $P(s)$, $W_S(s)$ *and* $W_T(s)$ *all proper transfer functions. Show that using* *state-space representation (with similar notations as in the previous case), the augmented plant is given by:*

$$G(s) = \left[\begin{array}{ccc|cc} A_P & 0 & 0 & 0 & B_P \\ -B_S C_P & A_S & 0 & B_S & -B_S D_P \\ B_T C_P & 0 & A_T & 0 & B_T D_P \\ \hline -D_S C_P & C_S & 0 & D_S & D_S D_P \\ D_T C_P & 0 & C_T & 0 & D_T D_P \\ -C_P & 0 & 0 & I & D_P \end{array}\right] \tag{6.108}$$

In general case (considering W_S, W_Q and W_T) we obtain:

$$G(s) = \left[\begin{array}{c|c} W_S & -W_S P \\ 0 & W_Q \\ 0 & W_T P \\ \hline I & -P \end{array}\right] \tag{6.109}$$

The state-space representation for this general case has a more sophisticated structure, so it is omitted here but can be obtained easily using modern software tools.

If the augmented plant is used for robust controller synthesis using the Robust Control Toolbox of Matlab, then it is essential that the weights W_S, W_Q and W_T be selected so that the augmented plant has a D_{12} matrix of full

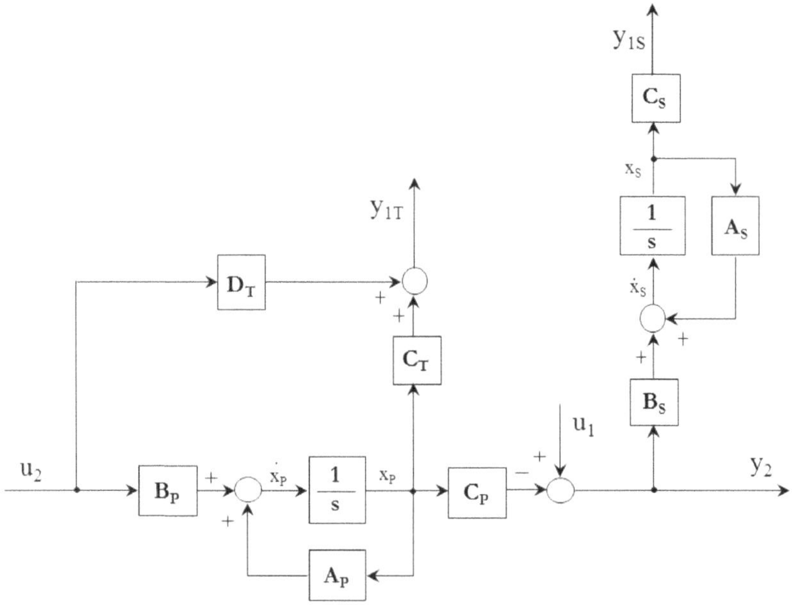

Fig. 6.18 Augmented plant using state-space representation.

column rank. An easy way to ensure this condition is to choose W_Q with an invertible $D-matrix$, for example $W_Q = \epsilon I$, where ϵ is any non-zero number.

6.7.2 Mixed-Sensitivity Optimization for Disturbance Rejection

Consider the unity feedback loop with two inputs d disturbance and n noise as well as two outputs e error and u control signal as shown in Fig. 6.19.

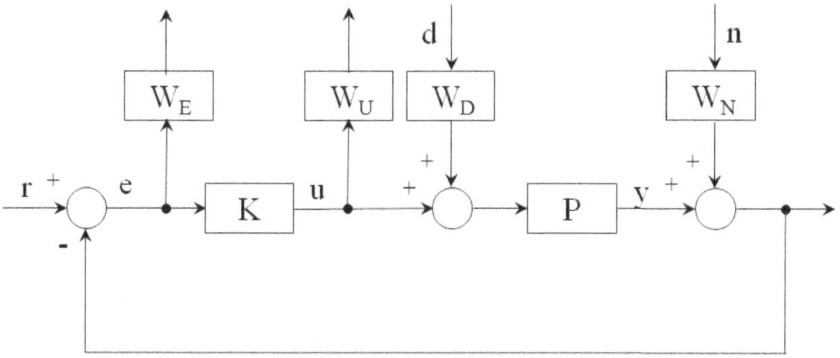

Fig. 6.19 Unity feedback loop with two inputs and two ouputs.

First let us consider that there are no weights on the inputs and outputs, then the transfer function in matrix form is:

$$\begin{bmatrix} e \\ u \end{bmatrix} = - \begin{bmatrix} PS & S \\ T & KS \end{bmatrix} \begin{bmatrix} d \\ n \end{bmatrix} \tag{6.110}$$

If we consider the weights as shown Fig. 6.19, we can write:

$$\begin{bmatrix} e \\ u \end{bmatrix} = - \begin{bmatrix} W_E & 0 \\ 0 & W_U \end{bmatrix} \begin{bmatrix} PS & S \\ T & KS \end{bmatrix} \begin{bmatrix} W_D & 0 \\ 0 & W_N \end{bmatrix} \begin{bmatrix} d \\ n \end{bmatrix} \tag{6.111}$$

Then the transfer function becomes:

$$T_{y1u1} = \begin{bmatrix} W_E & W_U \\ W_E & W_U \end{bmatrix} \begin{bmatrix} W_D P & W_N \\ W_D & W_N P^{-1} \end{bmatrix} \begin{bmatrix} S \\ T \end{bmatrix} \tag{6.112}$$

If we want to minimize T_{y1u1} in some sense we can specify:

$$\| |W_S S|^2 + |W_T T|^2 \|_\infty^{1/2} < 1 \tag{6.113}$$

where

$$|W_S| = |W_E| (|W_D P|^2 + |W_N|^2)^{1/2} \tag{6.114}$$

and

$$|W_T| = |W_U| (|W_D|^2 + |W_N|^2 |P|^{-2})^{1/2} \tag{6.115}$$

A particular case to the situation above is the disturbance rejection problem presented in Fig. 6.20 with:

$$|W_S| = |W_P|(|P|^2 + 1)^{1/2} \tag{6.116}$$

since $W_E = W_P$, $W_D = 1$ and $W_N = 1$.

Furthermore,

$$|W_T| = |W_U|(1 + |P|^{-2})^{1/2} \tag{6.117}$$

since $W_D = 1$ and $W_N = 1$.

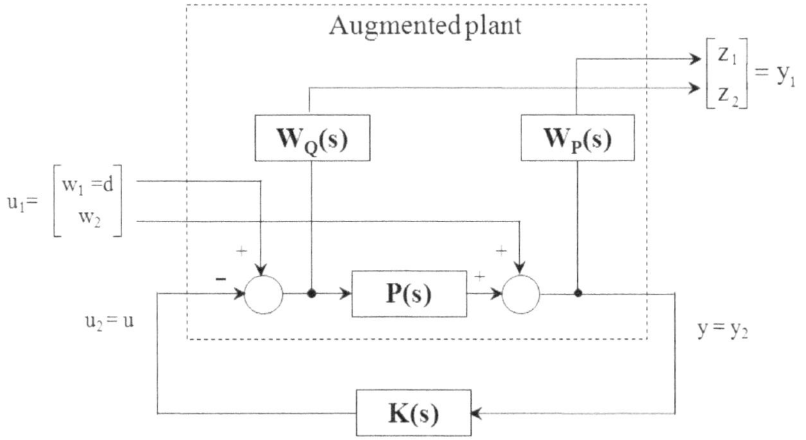

Fig. 6.20 Disturbance rejection problem.

The above problem can be re-formulated in terms of sensitivity S and control sensitivity transfer function $Q = KS$, so we would like to achieve:

$$\| |W_S S|^2 + |W_Q Q|^2 \|_\infty^{1/2} < 1 \tag{6.118}$$

Then in a similar way we can obtain:

$$|W_S| = |W_P|(|P|^2 + 1)^{1/2} \approx |W_P||P| \tag{6.119}$$

since $W_E = W_P$, $W_D = 1$ and $W_N = 1$. The above approximation is valid in the frequency range, where: $|P| >> 1$. Furthermore,

$$|W_Q| = |W_U||P|(1 + |P|^{-2})^{1/2} \approx |W_U| \tag{6.120}$$

since $W_D = 1$ and $W_N = 1$.

The above approximation is valid in the frequency range, where: $|P| >> 1$.

We remark W_Q corresponds to an additive uncertainty model.

In case of disturbance rejection control problem the design objective is to minimize the transfer function $T_{yd}(s)$ from the input disturbance d to the control system output y, written as:

$$T_{yd}(s) = \frac{P(s)}{1 + K(s)P(s)} \leq \gamma \tag{6.121}$$

or in a more general way:

$$T_{yd}(s) = \frac{P(s)}{1 + K(s)P(s)} \leq W_P(s) \tag{6.122}$$

where $W_P(s)$ is a frequency dependent performance weighting function.

The plant is augmented - as shown in Fig. 6.20 - with the additive model uncertainty W_Q and performance weighting function $W_P(s)$. Using state-space representation, we can write the augmented plant (left to the reader) in the following form:

$$\begin{bmatrix} \dot{x} \\ \dot{x}_{WP} \\ \dot{x}_{WQ} \\ z_1 \\ z_2 \\ y \end{bmatrix} = \left[\begin{array}{ccc|ccc} A & 0 & 0 & B & 0 & -B \\ B_{WP}C & A_{WP} & 0 & B_{WP}D & B_{WP} & -B_{WP}D \\ 0 & 0 & A_{WQ} & B_{WQ} & 0 & -B_{WQ} \\ \hline D_{WP}C & C_{WP} & 0 & D_{WP}D & D_{WP} & -D_{WP}D \\ 0 & 0 & C_{WQ} & D_{WQ} & 0 & -D_{WQ} \\ C & 0 & 0 & D & 1 & -D \end{array}\right] \begin{bmatrix} x \\ x_{WP} \\ x_{WQ} \\ w_1 \\ w_2 \\ u \end{bmatrix} \tag{6.123}$$

Chapter 7

Robust Controller Design

'A good decision is based on knowledge and not on numbers.' —Plato (424 B.C. - 348 B.C.)

Summary:

This section is dedicated to robust controller design. The design procedure is illustrated through examples. The following issues are discussed in this chapter:

- robust controller design using Youla parametrization
- robust controller design using Robust Control Toolbox of Matlab
- robust controller design with constraint on the control signal
- robust gain-scheduled control design and stability analysis.

7.1 Controller Design Using Youla Parametrization

An immediate application of the Youla parametrization is the robust performance problem. For a given stable plant P, with performance weight W_S and robustness weight W_T, we want to find a controller K in the form:

$$K = \frac{Q}{1 - PQ} \tag{7.1}$$

where Q is stable, such that:

$$\| |W_S(s)S(s)| + |W_T(s)T(s)| \|_\infty < 1 \tag{7.2}$$

In terms of the free parameter $Q(s)$, the sensitivity and complementary sensitivity functions are:

$$\begin{aligned} S(s) &= 1 - P(s)Q(s) \\ T(s) &= 1 - S(s) = P(s)Q(s) \end{aligned} \tag{7.3}$$

Hence, for robust performance, we need to design a proper stable $Q(s)$ satisfying the following inequality for all ω:

$$|W_S(j\omega)(1 - P(j\omega)Q(j\omega))| + |W_T(j\omega)P(j\omega)Q(j\omega)| < 1 \tag{7.4}$$

Next, let us focus only on minimum phase plants. In the light of the previous inequality, we design the controller by constructing a proper stable $Q(s)$ such that:

- $Q(j\omega) \approx P^{-1}(j\omega)$ when $|W_S(s)|$ is large and $|W_T(s)|$ is small;
- $Q(j\omega) \approx 0$ when $|W_S(s)| \approx 0$ (small) and $|W_T(s)|$ is large.

Obviously, if $W_S(s)$ and $W_T(s)$ are both large, it is impossible to achieve robust performance. In practice, $W_S(s)$ is large at low frequencies (for good tracking of reference inputs) and $W_T(s)$ is large at high frequencies for robustness against unmodelled dynamics.

Example 7.1. *Consider the following problem [61]:*

$$P(s) = \frac{2}{s^2 + 2s + 4}; W_s(s) = \frac{1}{s}; W_T(s) = \frac{s(1 + 0.1s)}{10} \tag{7.5}$$

Let $Q(s)$ be in the form:

$$Q(s) = P^{-1}\tilde{Q}(s) = \frac{s^2 + 2s + 4}{2}\tilde{Q}(s) \tag{7.6}$$

where $\tilde{Q}(s)$ is a proper stable transfer function whose relative degree is at least two. The robust performance design condition is equivalent to:

$$\Phi(\omega) = \left|\frac{1 - \tilde{Q}(j\omega)}{j\omega}\right| + \left|\frac{j\omega(1 + 0.1j\omega)}{10}\tilde{Q}(j\omega)\right| < 1, \forall\omega \tag{7.7}$$

So we should try to select:

$$\begin{aligned} &\tilde{Q}(j\omega) \approx 1 \quad \forall\omega < \omega_{low} \\ &\tilde{Q}(j\omega) << 1 \; \forall\omega > \omega_{high} \end{aligned} \tag{7.8}$$

where ω_{low} can be taken as the bandwidth of the filter $W_S(s)$, i.e. $\omega_{low} = \sqrt{2}rad/sec$ and ω_{high} can be taken as the bandwidth of the filter $W_T^{-1}(s)$, i.e. $\omega_{high} = 10rad/sec$.

Following these guidelines, let us choose:

$$\tilde{Q}(s) = \frac{1}{(1 + \tau s)^2} \tag{7.9}$$

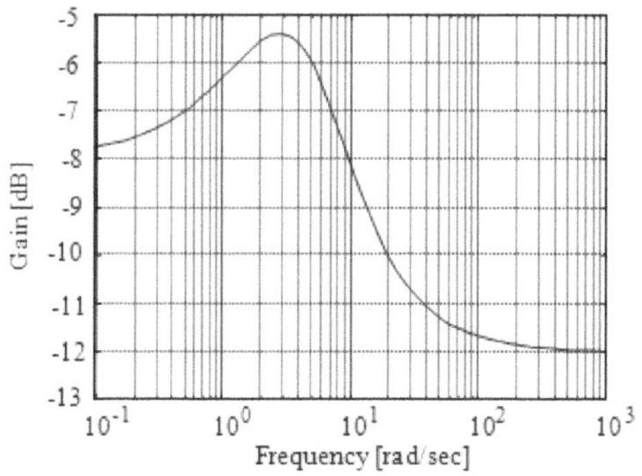

Fig. 7.1 Robust performance - $\Phi(\omega)$.

with τ^{-1} being close to the midpoint of the interval $(\sqrt{2}, 10)$, let's say $\tau = 1/5$. Then $Q(s)$ transfer function is given by:

$$Q(s) = \frac{s^2 + 2s + 4}{2(1 + 0.2s)^2} \tag{7.10}$$

As shown in Fig. 7.1 the $\Phi(\omega) < 1$, for all ω, therefore the robust performance condition is satisfied. The controller according with the Youla parametrization is:

$$K(s) = \frac{Q(s)}{1 - P(s)Q(s)} = \frac{5(s^2 + 2s + 4)}{2s(0.2s + 1)} \tag{7.11}$$

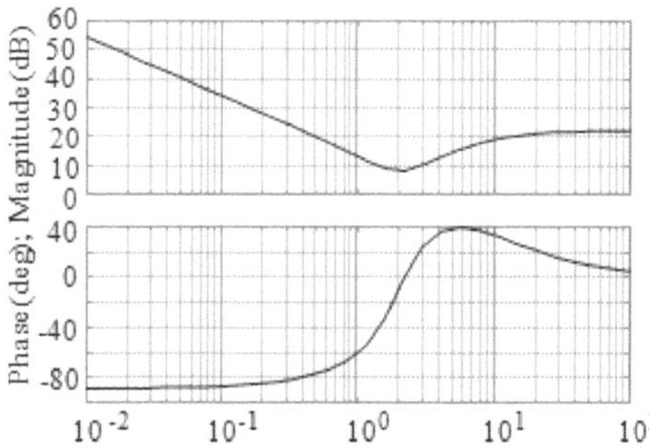

Fig. 7.2 Controller's Bode diagram.

The controller's Bode diagram is shown in Fig. 7.2, the system gain and phase margins are $G_m = 2.46 \cdot 10^5$ and $\phi_m = 76°$, presented in Fig. 7.3 as well as the step response is shown in Fig. 7.4.

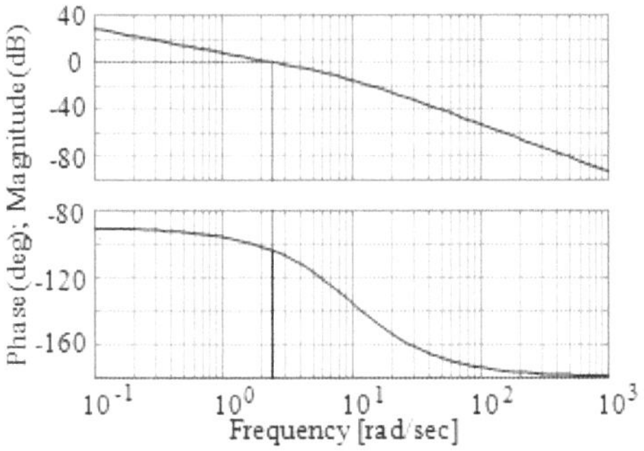

Fig. 7.3 Gain and phase margins.

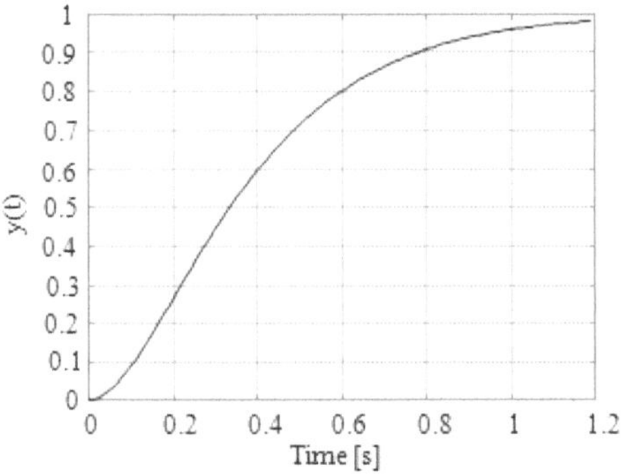

Fig. 7.4 Step response.

7.2 Controller Design Using Robust Control Toolbox

In this section the robust controller design is illustrated by using Robust Control Toolbox of Matlab.

We have seen in the previous chapter that the robust performance condition in case of multiplicative uncertainty is:

$$|W_S S| + |W_T T| < 1 \tag{7.12}$$

or

$$\| \mid W_S S \mid + \mid W_T T \mid \|_\infty < 1 \tag{7.13}$$

The transfer functions $P(s), W_S(s)$ and $W_T \cdot P(s)$ (in the framework of Matlab's Robust Control Toolbox) must be proper, i.e. bounded as $s \to \infty$, however W_T may be improper.

The default choice in the Matlab's Robust Control Toolbox is to scale all the output channels, usually we will scale only one output channel related to the sensitivity weighting function $W_S(s)$. The other weighting functions is related to plant uncertainty and are well-defined values in case of a given real plant.

The main design steps are presented via an example below.

Example 7.2. *Consider the following problem data:*

$$P(s) = \frac{1}{(s+1)(s+2)}; W_T(s) = \frac{s(1+0.1s)}{10} \tag{7.14}$$

and required performances are: $T_{set} = 1.2s$ *and* $P.O. = 4\%$.

First, the time domain performance specifications are translated into frequency domain: $\xi = 0.707$ and $\omega_n = 4/(\xi T_{set}) \approx 4.71$, which leads to:

$$W_S(s) = \frac{s^2 + 2\xi\omega_n s + \omega_n^2}{s(s + 2\xi\omega_n)} \approx \frac{s^2 + 2\xi\omega_n s + \omega_n^2}{(s + 0.001)(s + 2\xi\omega_n)} \tag{7.15}$$

The next step is to augment the plant considering the weighting functions $W_S(s)$ and $W_T(s)$. During γ iterations only the sensitivity weighting function $W_S(s)$ is scaled (iterated).

Solution obtained using the Robust Control Toolbox

The robust controller synthesis is presented in a Matlab session 7.2.

MATLAB session 7.2: Robust controller design.

```
% plant state-space representation
>> [ag,bg,cg,dg] = tf2ss(num,den);
% Specific to Robust Control Toolbox
>> ss_g = mksys(ag,bg,cg,dg);

% plant augmentation - WS, WQ, WT weighting functions
>> TSS_ = augtf(ss_g,WS,WQ,WT);
>> [a_p,b_p,c_p,d_p] = branch(TSS_);

% controller synthesis
>> [ss_f,ss_cl] = hinf(TSS_);

% optimal controller synthesis
% iterations on WS weighting function
>> [rho_opt, ss_f, ss_cl] = hinfopt(TSS_,1);

% Find out the controller
>> [A_f,B_f,C_f,D_f] = branch(ss_f);
>> [num_cont, den_cont] = ss2tf(A_f,B_f,C_f,D_f);
```

Table 7.1 $\mathcal{H}_\infty$ optimal control synthesis.

No	Gamma	D11 $\leq$ 1	P-Exist	P $\geq$ 0	S-Exist	S $\geq$ 0	lam(PS) < 1	C.L.
1	1.0000e+000	FAIL	FAIL	OK	FAIL	OK	OK	STAB
2	5.0000e-001	OK	OK	OK	OK	OK	OK	STAB
3	7.5000e-001	OK	OK	OK	OK	OK	OK	STAB
4	8.7500e-001	OK	OK	OK	OK	OK	OK	STAB
5	9.3750e-001	OK	OK	OK	OK	OK	OK	STAB
6	9.6875e-001	OK	OK	OK	OK	OK	OK	STAB
7	9.8438e-001	OK	OK	OK	OK	OK	OK	STAB
8	9.9219e-001	OK	OK	OK	OK	OK	OK	STAB

Iteration no. 8 is your best answer under the tolerance: 0.0100.

In this case the synthesized robust controller for the investigated second-order plant is:

$$K(s) = 3277 \frac{(s - z_1)(s - z_2)(s - z_3)}{(s - p_1)(s - p_2)(s - p_3)(s - p_4)} \tag{7.16}$$

with

$[z_1 \; z_2 \; z_3] = [-67.8398 \; -2.0 \; -1.0]$
$[p_{1,2} \; p_3 \; p_4] = [-72.2 \pm 69.05j \; -6.66 \; -0.001]$

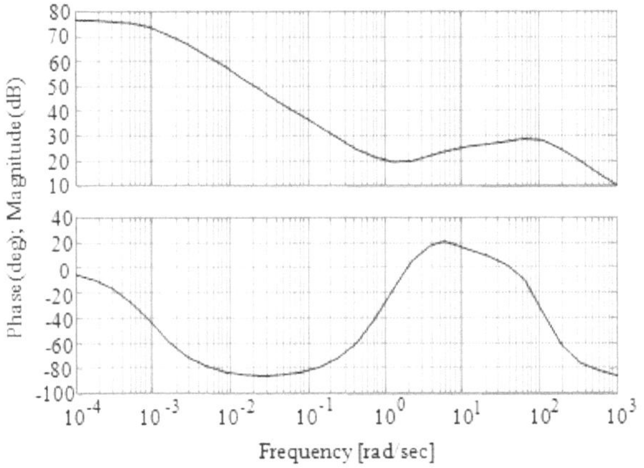

Fig. 7.5 Controller's Bode diagram.

The results of gamma iterations are shown in Table 7.1. The controller's Bode diagram is shown in Fig. 7.5, the system gain and phase margins are $G_m = 93.4$ and $Pm = 65.4°$, and the step response is shown in Fig. 7.6.

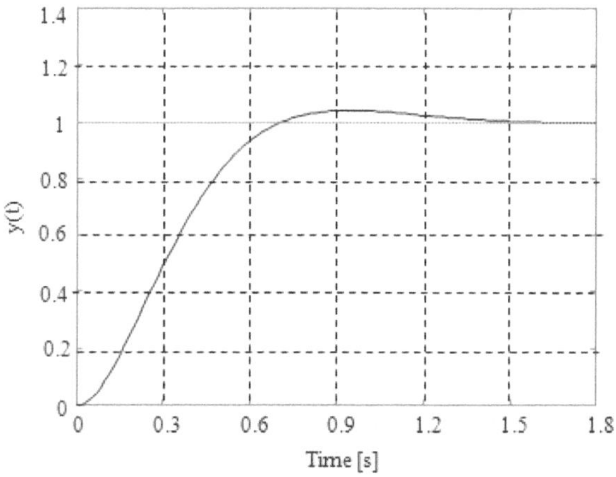

Fig. 7.6 Step response.

7.2.1 Robust Controller Design for Brushless D.C. Drive

The plant transfer function and the related multiplicative uncertainty model have been deduced using system identification:

$$P(s) = \frac{1}{20(s+2.5)(s+800)} \tag{7.17}$$

$$W_T(s) = 0.1\frac{s^2+400s+200^2}{200^2} \tag{7.18}$$

Furthermore, let us define the design specifications in terms of:

$$T_{set1} = 0.055s \; and \; P.O._1 = 4\% \tag{7.19}$$

where the index 1 refers to the controller $K_1(s)$ which we are going to design.

In the case of a second-order system, with damping ratio ξ and natural frequency ω_n, the settling time T_{set} and the percent of overshoot $P.O.$ can be expressed approximately as:

$$T_{set} \approx \frac{4}{\xi\omega_n} \tag{7.20}$$

$$P.O. = 100 \cdot e^{-\xi\pi\sqrt{1-\xi^2}} \tag{7.21}$$

From equation (7.20) and (7.21) it's possible to compute the damping and the natural frequency. The corresponding weighting function $W_{S1}(s)$ is defined according with the general equation (7.22) (for details see previous chapter).

$$W_S(s) = \frac{s^2 + 2\xi\omega_n s + \omega_n^2}{(s+0.001)(s+2\xi\omega_n)} \tag{7.22}$$

In this case the synthesized robust controller using the Robust Control Toolbox of Matlab is:

$$K_1(s) = 160.63\frac{(s-z_1)(s-z_2)(s-z_3)}{(s-p_1)(s-p_2)(s-p_3)(s-p_4)} \tag{7.23}$$

with
$[z_1\ z_2\ z_3] = [-1.38\ -0.8\ -0.0025] \cdot 10^3$
$[p_{1,2}\ p_3\ p_4] = [-1.5 \pm 1.41j\ -0.15\ -10^{-6}] \cdot 10^3$

The results of gamma iterations for the considered brushless d.c. drive are presented in Table 7.2, the obtained value for γ is less than 1 that means the robust performance condition is not satisfied. Despite this, we will adopt this design, since $\gamma = 0.992$ is very close to 1.

This result ($\gamma < 1$) was expected because we derived our performance weighting function for the ideal case - no model uncertainty has been considered. In order to satisfy the robust performance we have to relax the performance weighting function $W_{S1}(s)$.

Table 7.2 $\mathcal{H}_\infty$ optimal control synthesis - BLDC.

No	Gamma	D11 $\leq$ 1	P-Exist	P $\geq$ 0	S-Exist	S $\geq$ 0	lam(PS) < 1	C.L.
1	1.0000e+000	FAIL	FAIL	OK	FAIL	OK	OK	STAB
2	5.0000e-001	OK	OK	OK	OK	OK	OK	STAB
3	7.5000e-001	OK	OK	OK	OK	OK	OK	STAB
4	8.7500e-001	OK	OK	OK	OK	OK	OK	STAB
5	9.3750e-001	OK	OK	OK	OK	OK	OK	STAB
6	9.6875e-001	OK	OK	OK	OK	OK	OK	STAB
7	9.8438e-001	OK	OK	OK	OK	OK	OK	STAB
8	9.9219e-001	OK	OK	OK	OK	OK	OK	STAB

The Bode plot of the designed robust controller $K_1(s)$ is presented in Fig. 7.7 - curve 1, solid line.

Next, let us assume that the design does not fulfill the specification related to the control signal amplitude (it exceeds the specified limit).

In order to reduce the control amplitude (shown in Fig. 7.8 - curve 1, solid line), the simplest approach is to relax the performance specifications (increase the settling time).

Let us require:

$$T_{set2} = 0.3s \; and \; P.O._2 = 4\% \tag{7.24}$$

which will lead to the weighting function $W_{S2}(s)$ defined according with the general equation (7.22). Practically we reduce the gain of the controller, which

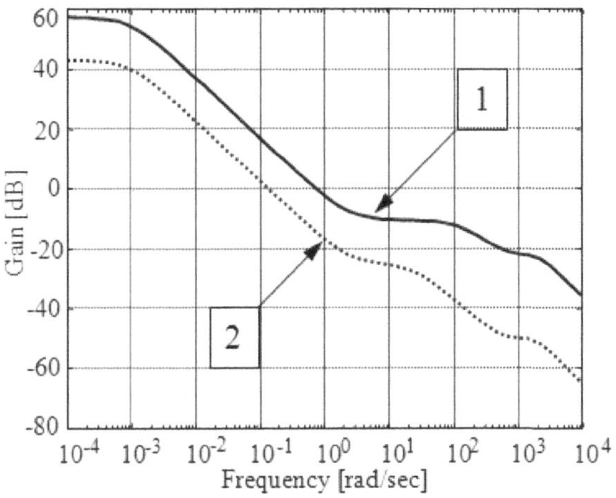

Fig. 7.7 Controller's Bode diagram.

leads to lower performance and smaller $Q(s)$ transfer function. It is left to the reader to verify, that $Q_2(s) < Q_1(s)$.

The synthesized controller has the following transfer function equation (7.25) as well as the related Bode plot is shown in Fig. 7.7 - curve 2, dotted line.

$$K_2(s) = 5.5876 \frac{(s - z_1)(s - z_2)(s - z_3)}{(s - p_1)(s - p_2)(s - p_3)(s - p_4)} \tag{7.25}$$

with

$[z_1\ z_2\ z_3] = [-1.03\ -0.8\ -0.0025] \cdot 10^3$

$[p_{1,2}\ p_3\ p_4] = [-1.28 \pm 1.26i\ -0.027\ -10^{-6}] \cdot 10^3$

The variation of the corresponding control signal, when the system is controlled by the $K_2(s)$ controller is presented in Fig. 7.8, curve 2 (dotted line). The system response is shown in Fig. 7.9, curve 2 (dotted line).

Limiting the control signal (authority) in the case of the $K_1(s)$ controller, by a simple saturation block has the following disadvantages:

- the system will manifest a larger overshoot and a longer settling time, than in the case of the $K_2(s)$ controller
- due to the saturation, the system will become a nonlinear system, therefore robustness cannot be guaranteed in a linear design framework.

The obtained experimental results - in the case of saturation - are presented in Fig. 7.8 (current response - curve 3, dashed line) and Fig. 7.9 (speed response - curve 3, dashed line). Later on in this chapter we will present a theoretical framework, which can be used to design robust controllers with limited control authority.

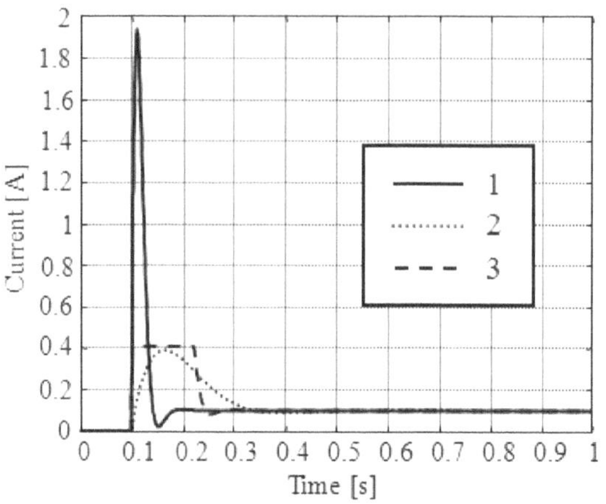

Fig. 7.8 The current response.

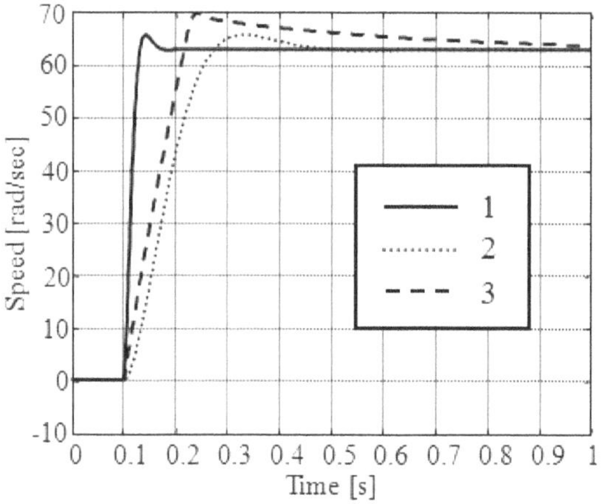

Fig. 7.9 Speed response.

7.2.2 Robust Controller Design for Active Vibration Suppression

In this section we consider a practical example, active vibration suppression of a three-story flexible structure.

The Mathematical Model of the Flexible Structure

The equation of motion for an idealized three-story flexible structure can be written in the following form [19]:

$$M\ddot{q} + C\dot{q} + Kq = m_a F_s \ddot{q}_a + M E_s \ddot{q}_g \tag{7.26}$$

$$M = diag\begin{bmatrix} m_1 \; m_2 \; m_3 \end{bmatrix}$$

$$C = \begin{bmatrix} c_1 + c_2 & -c_2 & 0 \\ -c_2 & c_2 + c_3 & -c_3 \\ 0 & -c_3 & c_3 \end{bmatrix}$$

$$K = \begin{bmatrix} k_1 + k_2 & -k_2 & 0 \\ -k_2 & k_2 + k_3 & -k_3 \\ 0 & -k_3 & k_3 \end{bmatrix}$$

$$F_s = \begin{bmatrix} 0 \; 0 \; -1 \end{bmatrix}^T \; E_s = \begin{bmatrix} -1 \; -1 \; -1 \end{bmatrix}^T \; q = \begin{bmatrix} q_1 \; q_2 \; q_3 \end{bmatrix}^T$$

The mass, damping and stiffness matrices of the structure (see Fig. 7.10) are noted by M, C and K, the relative displacement vector is noted by q. The first term on the right-hand side $m_a F_s \ddot{q}_a$ in equation (7.26) is the control force generated by the active mass damper using an auxiliary mass. The second term $M E_s \ddot{q}_g$ is the force disturbance.

The nominal parameters of the experimental set-up derived from material characteristics as well as the transfer gains: K_{gap} - transfer gain of the displacement sensor, K_{acc} - transfer gain of the acceleration sensor, K_{AMD} - transfer gain of motor drive, are presented in Table 7.3.

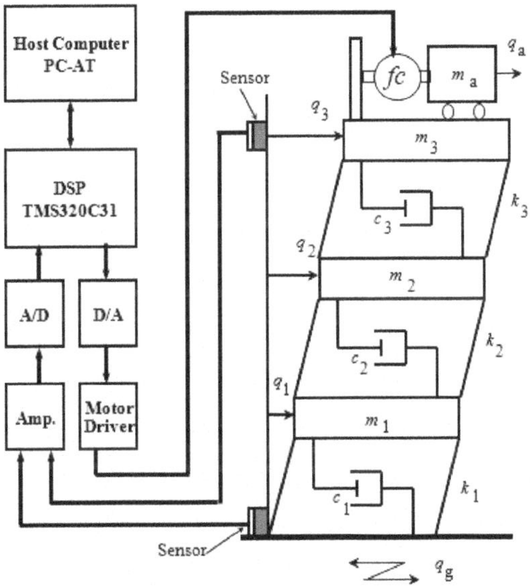

Fig. 7.10 Flexible structure.

Table 7.3 The nominal parameters.

m_1, m_2, m_3	1.74[kg], 1.65[kg], 2.34[kg]
m_a	0.047[kg]
c_1, c_2, c_3	2.2[Ns/m]
k_1, k_2, k_3	9233 [N/m]
transfer gain of the displacement sensor	2.5[V]/12.5[mm]
transfer gain of the accel. sensor	1[V]/2[m/s^2]
transfer gain of motor drive	3[N]/5[V]

In the case of the investigated flexible structure with: $c_1 = c_2 = c_3 = c$ and $k_1 = k_2 = k_3 = k$ as well as with the adopted sign convention, the related transfer functions (using displacement sensors), between the input force applied at the 3rd story by the AMD and the output (the relative displacement) measured at each story are expressed as:

$$P_{13}(s) = \frac{(cs+k)^2}{\beta_6 s^6 + ... + \beta_0} \tag{7.27}$$

$$P_{23}(s) = \frac{(m_1 s^2 + 2cs + k)(cs+k)}{\beta_6 s^6 + ... + \beta_0} \tag{7.28}$$

$$P_{33}(s) = \frac{\alpha_4 s^4 + ... + \alpha_0}{\beta_6 s^6 + ... + \beta_0} \tag{7.29}$$

where:

$$\begin{array}{llll} \alpha_4 = & m_1 m_2 & \alpha_3 = & 2c(m_1+m_2) \\ \alpha_2 = & 2k(m_1+m_2)+3c^2 & \alpha_1 = & 6ck \\ \alpha_0 = & 3k^2 & & \end{array} \tag{7.30}$$

and

$$\begin{array}{llll} \beta_6 = & m_1 m_2 m_3 & \beta_5 = & c m_A \\ \beta_4 = & k m_A + c^2 m_B & \beta_3 = & c^3 + 2ck m_B \\ \beta_2 = & k^2 m_B + 3c^2 k & \beta_1 = & 3ck^2 \\ \beta_0 = & k^3 & & \end{array} \tag{7.31}$$

with

$$\begin{array}{ll} m_A = & m_1 m_2 + 2 m_1 m_3 + 2 m_2 m_3 \\ m_B = & m_1 + 2m_2 + 3m_3 \end{array} \tag{7.32}$$

Furthermore, equation (7.29) can be rewritten as:

$$P_{33}(s) = \sum_{i=1}^{3} \frac{\alpha_i \omega_i^2}{s^2 + 2\xi_i \omega_i s + \omega_i^2} \tag{7.33}$$

The previous equation suggests that the resonance frequencies of the flexible structure (poles of the transfer function) as well as the coefficients α_i uniquely determine the transfer function, the zeros are obtained automatically.

In this laboratory experimental set-up (see Fig. 7.10) we measured the displacement of each story using a displacement (inductive) sensor. In practice displacement sensors cannot be used, acceleration sensors must be used. Based on acceleration measurements we can calculate the relative speed and displacements among the stories.

Identification of the Flexible Structure

In the early '90s identification of flexible structures has been in the focus of the scientific community and a vast literature about this topic exist [9, 14, 24].

In the first phase the flexible structure is identified based on measured input/output data, considering an autoregressive exogenous (ARX) model [51], [52]. In the identification process as input signal the pseudo-random binary signal (PRBS) is used.

The input signal is generated by the active mass driver and the output is the relative displacement between the base and the 3rd story. The input and output signals are presented in Fig. 7.11 (left side and right side).

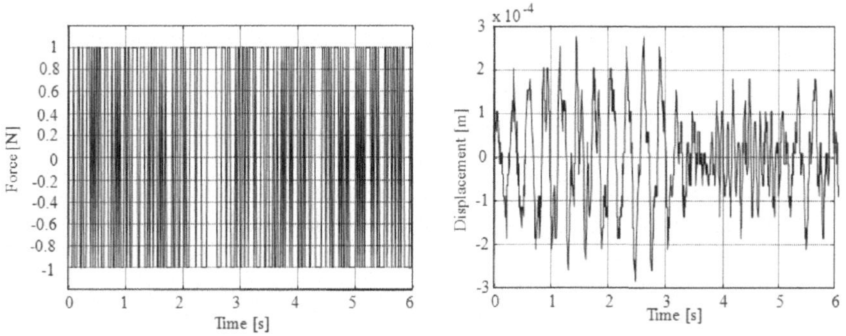

Fig. 7.11 The input and output signal.

Defining a frequency range for identification as ω_{id} the sampling time T_s is selected according to equation (7.34).

$$\frac{2\pi}{100T_s} < \omega_{id} < \frac{2\pi}{5T_s} \tag{7.34}$$

The main steps of the system identification and model-order reduction are described in chapter three, therefore are not repeated here.

We identify a high-order model (shown in Fig. 7.12 dotted line) and using model-order reduction methods (based on the gramiam) we and identify a six-order model (shown in Fig. 7.12 solid line) according to the physical model. The model is validated in time and frequency domain.

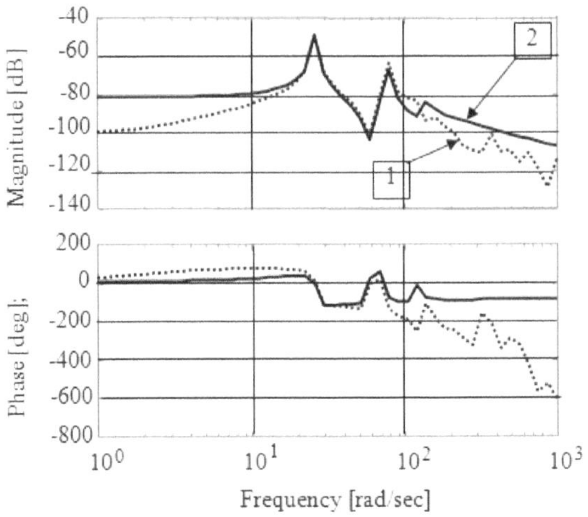

Fig. 7.12 Model validation in the frequency domain.

Robust Controller Design for Active Vibration Suppression

Active vibration suppression of flexible structures - in a robust control framework - has been the topic of several research papers [4, 5, 21, 48, 53], to cite just few of them.

The fundamental goal of many active and passive structural control strategies is to simply increase the natural damping of the flexible structure. A very simple way to increase the damping of a flexible structure is to use so-called rate feedback described in chapter 5.

A control strategy that simply inverts the plant (pole-zero cancellation) is not very effective/robust solution against structured parametric uncertainties (see variations of damping ratio and natural frequency).

In addition to structured parametric uncertainty, the robust controller design shall consider unstructured plant uncertainty due to the neglected high-frequency dynamics.

In the H_∞ control framework (see γ- iterations of Robust Control Toolbox of Matlab) the plant can be easily augmented considering unstructured plant uncertainty.

Although, parametric (structured) uncertainty can be translated into unstructured uncertainty it might lead to conservative design. The μ-synthesis framework (D-K iterations), which introduces a new norm (based on the structured singular value) is more suitable for handling parametric uncertainty, however discussion of this topic is out of the scope of the current book.

In the previous section the flexible structure has been identified and a 6th-order, model has been derived. This model might be used for control system design, however will lead to a relatively high-order controller. The industry is still reluctant to accept high-order controllers, due to difficulties in: understanding, real-time implementation, tuning, etc.

Therefore, a lower order model used in the controller design process seems a simpler and better approach. The difference between the high-order model and reduced order model, will be considered as plant uncertainty.

Let us assume that the nominal model is a second-order one (in this way is possible to keep the controller's order low), given by:

$$P(s) = \frac{K_{gain}}{2.34s^2 + s + 1500} \tag{7.35}$$

where $K_{gain} = 120[N/m]^{-1}$. This plant model includes the inductive sensor and actuator gains.

In the controller design process we will consider a multiplicative uncertainty model derived as:

$$\Delta_m(s) = \frac{\tilde{P}(s) - P(s)}{P(s)} \tag{7.36}$$

where $\tilde{P}$ is the real plant (high-order model) and $P(s)$ is the nominal plant (second-order model).

The gain plot related to the deduced multiplicative uncertainty is shown in Fig. 7.13, curve 2 - dotted line.

In order to take into account the model uncertainty, the weighting function $W_T(s)$ is made larger than the multiplicative uncertainty $\Delta_m(s)$ at any frequency [25].

$$|\Delta_m| \leq |W_T(j\omega)| \ \forall \omega \tag{7.37}$$

The transfer function $W_T(s)$, which satisfies the previous inequality (equation 7.37) might be chosen as (shown in Fig. 7.13, solid line):

$$W_T(s) = 0.5623\frac{s^2 + 84s + 60^2}{60^2} \tag{7.38}$$

It is verified that the zero-cross frequency of the weighting function $W_T(s)$ is larger than the first resonance frequency and the gain of the weighting function $W_T(s)$ is larger than that of Δ_m in the control frequency range.

In case of a disturbance rejection control problem (vibration suppression falls into this category) we might define the performance weighting function $W_S(s)$ starting from the desired (ideal) transfer function between disturbance $D(s)$ and output $Y(s)$ denoted by $M_{id}(s)$.

If the nominal plant transfer function is written as:

$$P(s) = K_{gain}\frac{\omega_n^2}{s^2 + 2\tau\omega_n s + \omega_n^2} \tag{7.39}$$

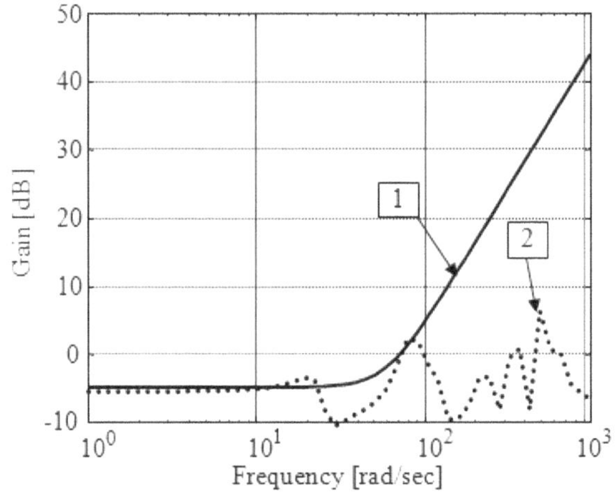

Fig. 7.13 Plant uncertainty - flexible structure.

where $\tau = 1.0215$ is the damping factor and $\omega_n = 25.3$ [rad/sec] is the natural frequency.

We might desire - via active vibration suppression - to increase the damping of the lightly damped poles. Therefore, the transfer function $M_{id}(s)$ can be written as:

$$M_{id}(s) = K_{gain} \frac{\omega_n^2}{s^2 + 2(\tau + \tau_a)\omega_n s + \omega_n^2} \tag{7.40}$$

where τ_a is the additional damping, which we want to introduce via active vibration suppression control.

However, we can write:

$$M_{id}(s) = \frac{P(s)}{1 + P(s)K(s)} = P(s)S_{id}(s) \tag{7.41}$$

where with $S_{id}(s)$ we denoted the ideal (desired) sensitivity transfer function.

Now it is possible to take the weighting transfer function $W_S(s)$ to be $S_{id}(s)^{-1}$, that is:

$$W_S(s) = \frac{P(s)}{M_{id}(s)} = \frac{s^2 + 2(\tau + \tau_a)\omega_n s + \omega_n^2}{s^2 + 2\tau s + \omega_n^2} \tag{7.42}$$

In our case we will chose $W_S(s)$ in the form:

$$W_S(s) = \gamma \frac{s^2 + 2(\tau + \tau_a)s + \omega_n^2}{s^2 + 2\tau s + \omega_n^2} \tag{7.43}$$

where γ is the parameter on which we iterate.

Therefore, we consider the performance weighting function as:

$$W_S(s) = 0.3\frac{2.34s^2 + 37s + 1500}{2.34s^2 + s + 1500} \tag{7.44}$$

where 0.3 is a scaling constant (to relax the ideal performance specifications).

The plant augmentation and the controller design is similar with the procedure presented in case of brushless d.c. drive, the γ-iterations are shown in Table. 7.4.

Table 7.4 $\mathcal{H}_\infty$ optimal control synthesis - flexible structure.

No	Gamma	D11 $\leq$ 1	P-Exist	P $\geq$ 0	S-Exist	S $\geq$ 0	lam(PS) $<$ 1	C.L.
1	1.0000e+000	OK	OK	OK	OK	OK	OK	STAB
2	2.0000e+000	OK	OK	OK	OK	OK	OK	STAB
3	4.0000e+000	FAIL	OK	OK	OK	OK	OK	STAB
4	3.0000e+000	OK	OK	FAIL	OK	OK	OK	UNST
5	2.5000e+000	OK	OK	OK	OK	OK	OK	STAB
6	2.7500e+000	OK	OK	OK	OK	OK	OK	STAB
7	2.8750e+000	OK	OK	OK	OK	OK	OK	STAB
8	2.9375e+000	OK	OK	OK	OK	OK	OK	STAB
9	2.9688e+000	OK	OK	OK	OK	OK	OK	STAB
10	2.9844e+000	OK	OK	FAIL	OK	OK	OK	UNST

The iteration no. 9 is the best answer under the tolerance: 0.0100 and the optimal γ is: $\gamma_{opt} = 2.9688$. Since $\gamma_{opt} > 1$ means robust stability and performance hold.

The controller's transfer function is:

$$K(s) = 267.97 \cdot 10^3 \frac{(s - 4.1829)}{(s + 177.2)(s + 4716.8)} \tag{7.45}$$

and its Bode plot is shown in Fig. 7.14.

Furthermore, the transfer function from input disturbance to output displacement is shown in Fig. 7.15, the solid line corresponds to the uncontrolled structure and the dashed line to the actively controlled structure.

Next, we can verify if the robust stability condition $\|W_T(s)T(s)\|_\infty < 1$ as well as the imposed performance condition $\|W_S(s)S(s)\|_\infty < 1$ hold. These plots are shown in Fig. 7.16.

We can also verify if the robust performance condition is fulfilled:

$$\| |W_S(s)S(s)| + |W_T(s)T(s)| \|_\infty < 1 \tag{7.46}$$

as shown in Fig. 7.17.

Finally, we can investigate the robustness of the system against structured parameter uncertainty, for example: $\pm 10\%$ variation of damping as well as $\pm 10\%$ of the variation of the natural frequency. Such a study, will reveal a good robustness of the designed controller.

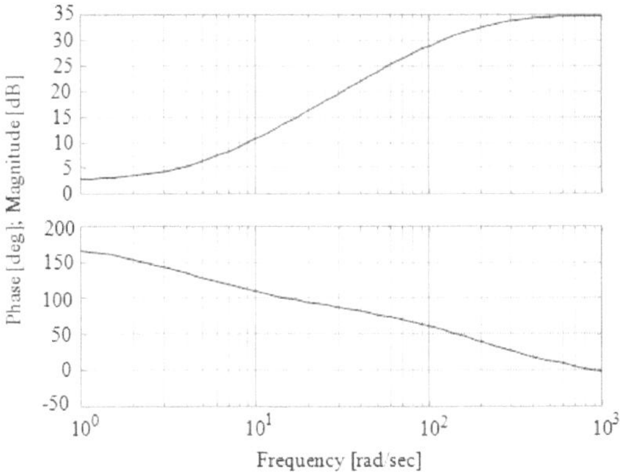

Fig. 7.14 Controller's Bode plot.

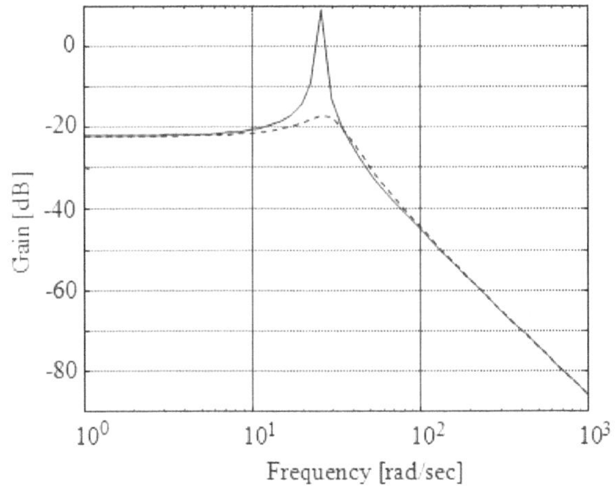

Fig. 7.15 Bode plots - controlled and uncontrolled nominal plant.

At the end of this section, we would like to mention some limitations of the H_∞ methods.

The H_∞-controller design method might lead to high-order controllers. A high-order controller is more difficult to implement in real-time than a low-order controller. It requires more hardware resources and might be computationally fragile. The order of the controller is the order of the plant and the sum of the orders of all weighting functions.

Therefore, in practice is important to find a good balance between performance, robustness and real-time implementation.

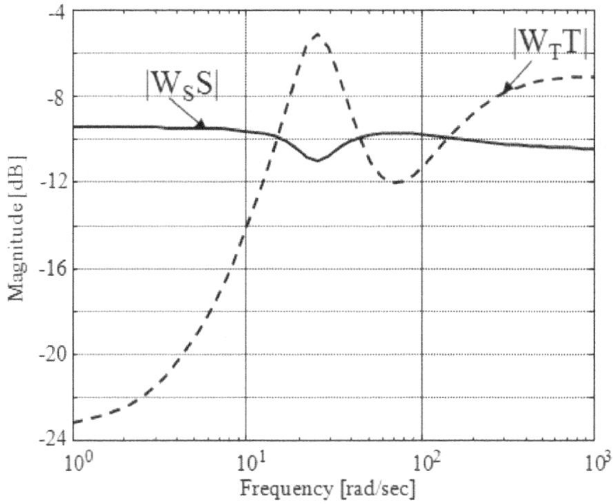

Fig. 7.16 Robust stability and performance conditions.

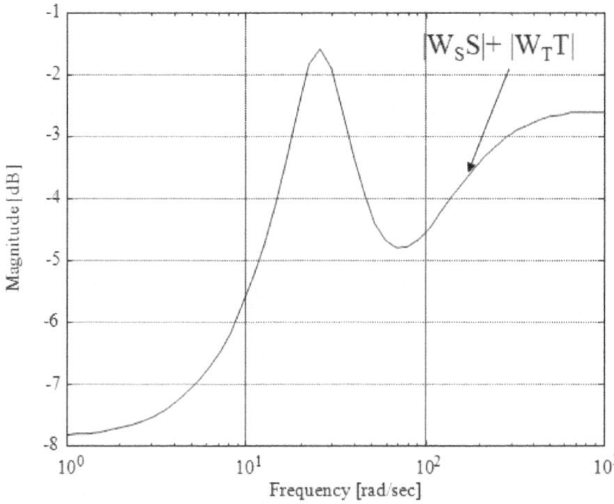

Fig. 7.17 Robust performance condition.

To achieve a low-order controller, the following measures might be applied:

- keep the order of the weighting functions as low as possible
- design the controller for a reduced order model, but include the neglected dynamics into the plant uncertainty
- design the controller for a nominal plant (high-order model) and after synthesis reduce the order of the controller.

Furthermore, the uncertainty bounds (additive or multiplicative), which we considered are not tight uncertainty bounds especially in case of structured plant uncertainty. These uncertainty bounds usually incorporates a much larger uncertainty bound than necessary and lead to a conservative controller design. This issue is handled in a better way, when a structured uncertainty is used in conjunction with μ synthesis.

7.3 Controller Design with Constraint on the Control Signal

Controller design for systems with hard constraints is a quite vivid area of research, due to the fact that most practical control problems are dominated by constraints on the control signal [67], [68]. Exceeding the prescribed bounds causes unexpected behavior of the system - large overshoot, low performance or (in the worst case) instability.

A general framework for the design of control systems subject to hard constraints is presented in [62]. The design procedure is based on the $\mathcal{H}_\infty$ loop shaping and relies on the calculation of the maximum control amplitude for a class of reference signals, bounded in amplitude and rate.

Controller designs that take into account the saturation effect a-priori are usually separated into two categories:

- designs that prevent saturation of the control signal and therefore enjoy a linear framework (as long as plant and controller are linear)
- methods that allow saturation and therefore facing a nonlinear setup.

We will focus on the first category - saturation avoiding - philosophy. To solve the constraint control problem, implicitly we have to restrict the amplitude as well as the rate of the external signals. In many practical situations, this is a more accurate description, than without rate restriction. For example, in the case of tank, not only the liquid-level is bounded (by the tank's height) in addition the liquid-level cannot change arbitrarily fast. During our investigation we will consider a SISO (Single-Input Single-Output) system, although the procedure can be extended for MIMO (Multi-Input Multi-Output) systems [63].

Definition 7.1. (Admissible Reference Signal) *Let $R, \dot{R} \in \Re$ with $R \geq 0$ and $\dot{R} \geq 0$. A continuous piecewise differentiable reference signal r, fulfilling $r(t) = 0$ for all $t \leq 0$ is called $(R, \dot{R})$ admissible, when the following properties hold:*

$$\begin{aligned} \mid r(t) \mid &\leq R \; \forall t > 0 \\ \mid \dot{r}(t) \mid &\leq \dot{R} \; \forall t > 0 \end{aligned} \tag{7.47}$$

The set of all $(R, \dot{R})$ - admissible reference signals - is denoted by $\mathcal{A}(R, \dot{R})$.

Definition 7.2. (Maximum Control Amplitude) *Given the internally stable standard control loop the maximum control amplitude is the ∞-norm of the control signal:*

$$u_{max} = \|u\|_\infty = \sup_t |u(t)|,\; r \in \mathcal{A}(R, \dot{R}) \tag{7.48}$$

Moreover, let us denote the transfer function from reference r to control signal u with:

$$Q(s) = \frac{K(s)}{1 + K(s)P(s)} \tag{7.49}$$

Theorem 7.1. (Glover 1987) *Let $Q(s)$ be asymptotically stable and strictly proper transfer function with McMillan degree n, input r, output u. Then the following inequality holds:*

$$\|u\|_\infty \leq 2n\|Q\|_\infty \|r\|_\infty \tag{7.50}$$

upper bound for maximum control amplitude [37].

As a remark, the complexity of linear time-invariant models is generally defined as the dimension of the state vector of any minimal state-space representation of the system. This is also known as the McMillan degree (order) of the system. Moreover, the previous theorem can be extended to the case of proper transfer functions, too [63].

According with the previous theorem an explicit relation exists between the ∞-norm of the transfer function $Q(s)$ (i.e. the maximum over all frequencies of the largest singular value), and the maximum control amplitude.

The next question is, if there exists a relation between the singular values of the weights and those of the transfer function $Q(s)$? This is stated in [58], that decreasing the weights decreases the transfer function $Q(s)$ (in terms of the largest singular value). Hence is observed that decreasing the ∞-norm of the weights decreases an upper bound for the maximum control amplitude.

Within the Loop Shaping, this has the following consequences:

- we have to decrease the maximum singular value of the design weights in the frequency range where the ∞-norm of $Q(s)$ transfer function is large
- in the case of too low maximum control signal, we have to increase the maximum singular value in that frequency range.

A loop shaping algorithm for controller design in the case of systems with constraints is described in [62], [63].

Algorithm 1. (Loop Shaping Design) *Given a plant $P(s)$, with restrictions on the reference signal and the desired control amplitude $u_{max} > 0$. The loop shaping might be performed in the following way:*

1. *choose the design weights $W_S(s)$ and $W_T(s)$*
2. *design the controller for the shaped plant and compute the stability margin*

3. *decide whether the design-objectives are fulfilled or not:*
   - *is the stability margin large enough*
   - *are the performance-objectives fulfilled*
   - *does* $u(t) \leq u_{max}$ *hold for all admissible reference signals* $(R, \dot{R})$.
4. *If not, adapt weight* $W_S(s)$ *and go to step 2.*

The previous algorithm can be applied for motion control systems too, in order to design a robust controller that prevent the saturation of the control signal.

As conclusion, in order to reduce the control amplitude we have to decrease the maximum singular value of the design weight $W_S(s)$ in the frequency range where, the ∞-norm of the transfer function $Q(s)$ is large. However, shaping the $W_S(s)$ transfer function might lead to a higher-order controller, which is not really appreciated in practice. A much simpler approach has been presented in the previous section, where the performance weighting function simply has been scaled, and a price for this simplicity was the lower performance.

Until now our study has been focused on linear-time invariant systems and the control design process led to controllers with fixed parameters.

In many practical applications, the plant is time varying or the parameters are varying with the operating point. When the parameters undergo large variations, it is often impossible to achieve high performance over entire operating range with a single linear time-invariant controller. In order to achieve high-performance in face of large parameter variations, controllers with adjustable parameters are used.

In the next section we are going to discuss robust gain-scheduled control. Gain-scheduled control is very useful for systems whose dynamical equations depend on physical coefficients that vary during operation. If the parameter values are measured in real time, it is then desirable to use controllers that incorporate such measurements to adjust to the current operating conditions - such controllers are scheduled by the parameter measurements. This control strategy typically achieves higher performance in the face of large variations in operating conditions.

7.4 Robust Gain-Scheduled Control

Gain scheduling is a widely used technique for controlling certain classes of nonlinear or linear time varying systems. Rather than seeking a single robust linear time-invariant (LTI) controller for the entire operating range, gain scheduling consists in designing an LTI controller for each operating point and in a switching controller when the operating conditions change.

The main features of gain scheduling can be summarized as:

- gain scheduling employs powerful linear design tools (well-understood) on difficult nonlinear problems
- performance specifications are formulated in linear terms

- gain scheduling enables a controller to respond rapidly to changing operating conditions.

In the aim to treat the control design in a unified framework we have to address three important issues:

- linearization of nonlinear systems and linear parameter-varying model derivation
- robust controller design for LTI systems
- robust gain-scheduled controller design and stability analysis.

In the framework of Matlab's Linear Matrix Inequalities (LMI) Toolbox is possible to design gain-scheduled $\mathcal{H}_\infty$ controllers for linear parameter-dependent systems. The available tools are applicable to time varying and/or nonlinear systems, with dynamics reasonably well approximated by parameter dependent models.

The designed robust gain-scheduled control system must satisfy the following properties:

- the closed-loop system is stable for all admissible parameters
- closed-loop system must satisfy the robust performance condition.

In fact, it is known that when gain scheduling is used the stability of the "frozen" system ensures the stability of the time varying system for very slow relative changes of the varying parameters. In the framework of LMI Toolbox it is possible to check the stability of the gain-scheduled system considering the variation range and the variation rate of the time varying parameter.

Before, to start to discuss this topic let us briefly recall that: a nonlinear system of the form:

$$\dot{x} = f(x) + \sum_{i=1}^{m} g_i(x)u_i \tag{7.51}$$

where x is the state vector, u_i are the system inputs. For smooth functions f and g, the system above is called control-affine system.

This kind of systems have been extensively studied in nonlinear control theory [46]. The systems are linear in the actions but nonlinear with respect to the state.

7.4.1 Theoretical Background

Gain scheduling is a powerful controller design method with increasing interest in the control community [50, 70, 66]. In board terms, the design of a gain-scheduled controller for a nonlinear plant can be described as a four-step procedure:

- The first step is to compute a linear parameter-varying model for the plant. The most common approach is linearization of the nonlinear plant around operating points or set points [46]. Therefore, a linear parameter-variant plant (LPV) is obtained such as:

$$\begin{aligned} E(p)\dot{x} &= A(p)x + B(p)u \\ y &= C(p)x + D(p)u \end{aligned} \tag{7.52}$$

where $A(p), B(p), C(p), D(p)$ and $E(p)$ are known functions of some parameter vector $p = [p_1, ..., p_n]$.

- The second step is to use linear design methods to design linear controllers for the linear parameter-varying plant model. This design process may result directly in a family of linear controllers, such that for each fixed value of the parameter, the linear closed-loop system exhibits desirable performance.

$$K(p) = \begin{bmatrix} A_c(p) & B_c(p) \\ C_c(p) & D_c(p) \end{bmatrix} \tag{7.53}$$

- The third step is the implementation of linear controllers such that the controller coefficients (gains) are varied (scheduled) according to the current value of the scheduling variables. As the plant $P(s)$ and the controller $K(s)$ depend on time varying parameter p, in general, we should solve a number of $\mathcal{H}_\infty$ control problems corresponding to each value of p and interpolate them. However, in the case of affine parameter dependent systems, we need to solve only the problems corresponding to the vertices. The simplest case is to consider only two vertices, then the controllers are interpolated as follows:

$$\begin{bmatrix} A_c(p) & B_c(p) \\ C_c(p) & D_c(p) \end{bmatrix} = \frac{p(t) - p_{min}}{p_{max} - p_{min}} \begin{bmatrix} A_c^{(1)} & B_c^{(1)} \\ C_c^{(1)} & D_c^{(1)} \end{bmatrix} + \frac{p_{max} - p(t)}{p_{max} - p_{min}} \begin{bmatrix} A_c^{(2)} & B_c^{(2)} \\ C_c^{(2)} & D_c^{(2)} \end{bmatrix} \tag{7.54}$$

to get a gain-scheduled controller. In the case of $p(t) = p_{max}$ we obtain the controller for vertex 1 and for $p(t) = p_{min}$ the controller for vertex 2. If the parameter vector $p(t)$ takes N_p different values (N_p dimensional parameter box) then based on convex decomposition the linear parameter dependent controller is derived from controllers defined at the corner of the parameter box by:

$$\begin{bmatrix} A_c(p) & B_c(p) \\ C_c(p) & D_c(p) \end{bmatrix} = \sum_{i=1}^{N_p} \alpha_i \begin{bmatrix} A_c^{(i)} & B_c^{(i)} \\ C_c^{(i)} & D_c^{(i)} \end{bmatrix} \tag{7.55}$$

where $\alpha_i \geq 0$ and $\sum_i \alpha_i = 1$.

During our investigations, we will consider a parameter box with only two dimensions, this leads to lower performance levels but allows simple real-time implementation.

- The fourth step is performance assessment. Typically, local stability and performance properties of the gain-scheduled controller might be subject to analytical investigation, while the non-local performance evaluation is based on simulation studies.

7.4.2 Stability Analysis

In the framework of Matlab's Linear Matrix Inequalities (LMI) Toolbox it is possible to check the quadratic stability as well as the robust stability (biquadratic stability) of the gain-scheduled system considering the variation range and the variation rate of the time varying parameter.

For affine parameter-dependent systems:

$$E(p)\dot{x} = A(p)x \; p(t) = [p_1(t), ..., p_n(t)] \tag{7.56}$$

Quadratic stability

In the case of quadratic stability analysis we seek a positive matrix $P_L > 0$, independent of uncertain parameters, such that the quadratic Lyapunov function $V(x) = x^T P_L x$ proves the stability of the system for all admissible parameters.

Since the Lyapunov function is parameter independent, stability holds even when the parameters change arbitrarily fast, which can be quite conservative in many applications.

The LMI Toolbox's `quadstab` function assesses quadratic stability by solving the following LMI problem:

Minimize τ over $Q = Q^T$ such that:

$$\begin{aligned} A^T QE + EQA^T < \tau I \; \forall (A, E) \\ Q > I \end{aligned} \tag{7.57}$$

The global minimum of this problem is returned in τ and the system is quadratically stable if $\tau < 0$. Given the solution Q_{opt} of the LMI optimization, the Lyapunov matrix P_L is given by $P_L = Q_{opt}^{-1}$.

Furthermore, the `quadstab` function computes the largest portion of the specified parameter range where quadratic stability holds. Specifically, if each parameter p_i varies in the interval: $p_i \in [p_{i0} - \delta_i, p_{i0} + \delta_i]$ `quadstab` computes the largest $\theta > 0$ such that quadratic stability holds over the parameter box: $p_i \in [p_{i0} - \theta\delta_i, p_{i0} + \theta\delta_i]$. The system is quadratically stable if the returned parameter $\theta > 1$, for details see [35].

Biquadratic stability

In the case of biquadratic stability analysis we seek a positive definite parameter-dependent Lyapunov function of the form $V(x, p) = x^T Q(p)^{-1} x$ where

$$Q(p) = Q_0 + p_1 Q_1 + ... + p_n Q_n \tag{7.58}$$

For such Lyapunov function, the stability condition $dV(x, p)/dt < 0$ is equivalent to:

$$E(p)Q(p)A^T(p) + A(p)Q(p)E(p)^T - E(p)\frac{dQ}{dt}E(p)^T < 0 \tag{7.59}$$

The parameter-dependent Lyapunov approach - `pdlstab` function of LMI Toolbox - takes the advantage of the existence of bounds on the variation rates of the parameters - it is less conservative test and is useful in the case of slowly varying parameter dependent systems.

As a remark, the gain scheduling control technique is based on the fundamental assumption that the values of the parameters are known on-line by the controller. In general, the lack of information related to the parameters may lead to a failure in achieving the desired specifications.

7.4.3 Robust Gain-Scheduled Control Design – Example

Example 7.3. *Let us consider a saturated electromagnet, described by linear parameter-dependent model:*

$$P(L, s) = \frac{1}{sL(i) + R} \tag{7.60}$$

where i is the current, $R = 6.8\Omega$ is the coil resistance and $L = L(i)$ is the coil inductance, which is current dependent. Measurements showed that at $i_1 = 0.2A$, $L_1 = 2H$ and at $i_2 = 2A$, $L_2 = 0.5H$ - as the current is increasing the magnetic core saturates.

Our goal is to control the current with high accuracy despite inductance variation. The design specifications are: settling time $T_{set} = 0.1s$ and percent of overshoot $P.O. < 1.5\%$.

Due to large parameter variations $L \in [L_{med} - \delta_L, L_{med} + \delta_L]$ with $L_{med} = 1.25H$ and $\delta_L = 0.75H$, it is difficult the handle the problem using a single linear robust controller. Therefore, we seek a set of linear controllers and a switching law among them.

The Bode plot of the linear parameter varying system, for different parameter values $L = L(i)$ is shown in Fig. 7.18.

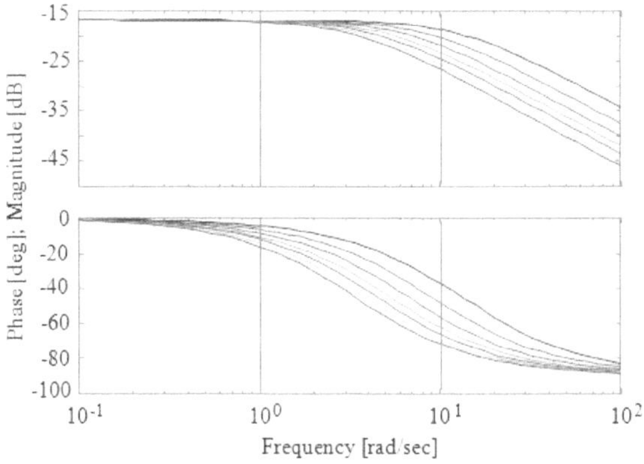

Fig. 7.18 Saturated electromagnet - Bode plot.

System identification experiments, performed around different operating points show that the parameter uncertainty around each operating point is bounded by:

$$W_T(s) = 0.1\frac{s+50}{50} \tag{7.61}$$

plant uncertainty.

The performance specifications given in time-domain are translated into frequency-domain, thus the performance weighting function is:

$$W_S(s) = \frac{s^2+80s+2500}{(s+0.0001)(s+80)} \tag{7.62}$$

We design for each operating point a robust controller and a switching controller, when the operating conditions change. We consider only two operating points around $i_1 = 0.2A, L_1 = 2H$ and $i_2 = 2A, L_2 = 0.5H$. The rate of change of the parameter L is bounded: $|dL/dt| < 15[H/s]$ since the current rate is bounded by $|di/dt| < 18[A/s]$, limitation basically imposed by the nonlinear inductance and power supply.

The parameter variation range and rate of change is shown in Fig. 7.19

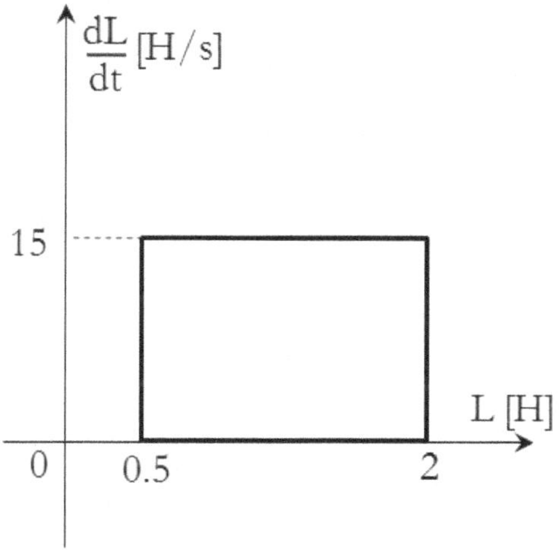

Fig. 7.19 Parameter variation range and rate of change.

The designed robust controllers for the two operating points are $K_1(s)$ and $K_2(s)$, whose transfer functions are given bellow:

$$K_1(s) = 1.863 \cdot 10^3 \frac{(s+13.6)(s+2721.1)}{(s+0.0001)(s+80)(s+4037)} \tag{7.63}$$

as well as

$$K_2(s) = 7.453 \cdot 10^3 \frac{(s+3.4)(s+2721.1)}{(s+0.0001)(s+80)(s+4037)} \tag{7.64}$$

The gain-scheduled controller is given by:

$$K(\alpha, s) = \alpha K_1(s) + (1-\alpha)K_2(s) \tag{7.65}$$

where

$$\begin{array}{l} \alpha = 0 \; K(\alpha, s) = K_1(s) \; i = 0.2A \\ \alpha = 1 \; K(\alpha, s) = K_2(s) \;\; i = 2A \end{array} \tag{7.66}$$

The varying parameter in this example is the inductance $L = L(i)$, but the scheduling parameter is the measured current i.

The previous equation can be written as:

$$K(L, s) = \frac{L-0.5}{1.5} K_1(s) + \frac{2-L}{1.5} K_2(s) \tag{7.67}$$

or in terms of the measured current i:

$$K(i, s) = \frac{i-0.2}{1.8} K_1(s) + \frac{2-i}{1.8} K_2(s) \tag{7.68}$$

The block diagram of the control system is shown in Fig. 7.20.

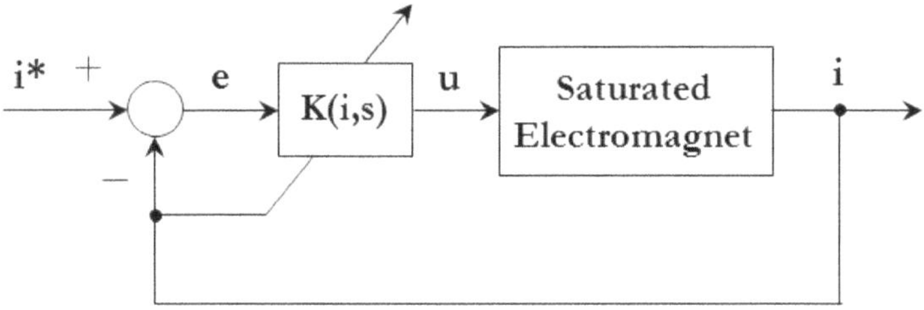

Fig. 7.20 Gain-scheduled control system.

The next step is to verify if the designed gain-scheduled control system is robust when the parameters are varying.

Therefore, biquadratic analysis is performed for the closed-loop parameter varying system using the LMI Toolbox of Matlab. The parameter L is varying in a parameter box as shown in Fig. 7.19.

The perfomed analysis shows that biquadratic stability holds for the considered range and rate of the time varying parameter.

Last, the step response of the robust gain-scheduled control system - for two different operating points - is presented in Fig. 7.21 and Fig. 7.22.

If we compare the the gain-scheduled control system step response (solid line) with the step response (dotted line) of linear time invariant (LTI) controllers ($K_1(s)$ and $K_2(s)$), we remark that using a single controller, performance cannot be assured in the entire operating range.

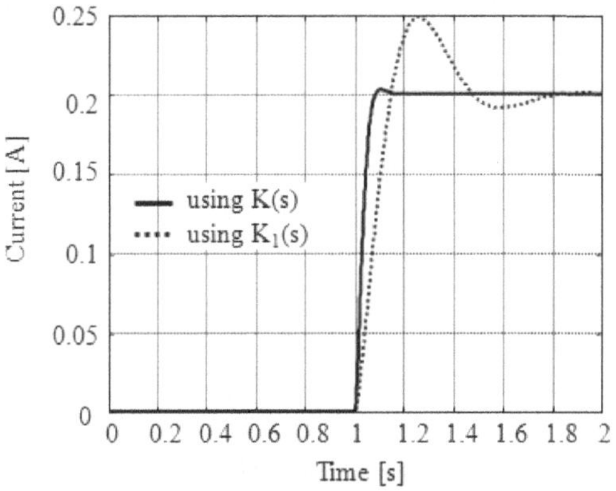

Fig. 7.21 Step response at 0.2 A.

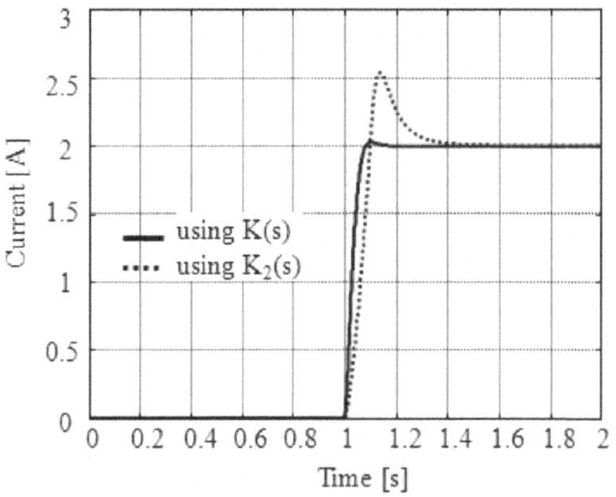

Fig. 7.22 Step response at 2 A.

In the considered example, the physical parameter (inductance) is dependent on the operating point (current), which can be easily measured. Therefore, high performance and robustness can be achieved with minimal cost by slightly increasing the complexity of the controller.

There are other practical examples where robust gain-scheduled control is successfully used such as: aircraft control (where the dynamics strongly depend on: angle of attack, speed and altitude) or car steering control, where the controller is desired to work well inside the operating domain defined by car speed and road conditions (friction coefficient).

7.5 Control Algorithm Implementation in Real-Time

Finally, the developed control algorithm shall be implemented in real-time.

The transfer function of the controller is discretized - for example - using the bilinear Tustin transformation, which conserve the ∞-norm:

$$s = \frac{2}{T_s} \frac{1 - z^{-1}}{1 + z^{-1}} \tag{7.69}$$

where T_s is the sampling time.

Nowadays, there are several integrated development environments where these steps are automated and the code is automatically generated. Furthermore, processing power has been significantly increased in the past years and floating-point operations are supported on different embedded platforms.

Despite these technological progresses there might be practical situations - cost constraints - where a fixed-point implementation is required.

Next, we will try to illustrate the real-time implementation of simple transfer functions using 16 bit fixed-point arithmetic.

Example 7.4. *Let assume that the transfer function of a low-pass filter is given by:*

$$LPF(s) = \frac{Y(S)}{U(s)} = \frac{500}{s + 500} \tag{7.70}$$

Implement this low-pass filter in real-time, in 16 bit fixed point arithmetic, assuming that the sampling time of the system is: $T_s = 0.001$ *seconds.*

The first step is to discretize the continuous transfer function given by (7.70) (using for example the Tustin transformation), when the sampling time is $T_s = 0.001$ seconds.

Thus we obtain:

$$\frac{Y(z)}{U(z)} = \frac{0.2z + 0.2}{z - 0.6} = \frac{0.2 + 0.2z^{-1}}{1 - 0.6z^{-1}} \tag{7.71}$$

The previous equation can be written in the following form:

$$y_k = 0.6y_{k-1} + 0.2(u_k + u_{k-1}) \tag{7.72}$$

where u_k and y_k is the input and output at the moment t_k and $t_k - t_{k-1} = T_s$.

Obviously, the previous equation in the given form, cannot be implemented in fixed-point arithmetic, therefore further transformations are required.

The low-pass filter can be implemented as:

$$y_k = \frac{614y_{k-1}}{1024} + \frac{205(u_k + u_{k-1})}{1024} \tag{7.73}$$

since the multiplication is supported by hardware and division by $1024 = 2^{10}$ is equivalent with a 10 bits right shift.

Even a complex controller, de-composed into elementary first-order transfer functions can be implemented in this way.

In the aim to get closer to a real problem, let us consider the following example.

Example 7.5. *Implement in 16 bit fixed-point arithmetic the following function:*

$$error = v - Ri \tag{7.74}$$

where v is the measured voltage, i is the measured current and R is the resistance. Furthermore, the related maximum values are: $v_{max} = 48V$, $i_{max} = 10A$, $R_{max} = 48\Omega$. In the aim to avoid numerical overflow as well as to assure numerical accuracy the variables v, i and R are scaled as follows:

$$v_{scale} = \frac{500[div]}{48[V]} \tag{7.75}$$

$$i_{scale} = \frac{2^{15}[div]}{10[A]} \tag{7.76}$$

$$R_{scale} = \frac{10000[div]}{48[\Omega]} \tag{7.77}$$

Obviously, in order to compute the correct value of the error according to equation (7.74), v as well as the product Ri must have the same scaling factors. Therefore, the error in fixed-point arithmetic is computed as:

$$error[div] = v - \alpha Ri \tag{7.78}$$

where the scaling factor α is given by:

$$\alpha = \frac{v_{scale}}{R_{scale} i_{scale}} = \frac{500}{48} \cdot \frac{10}{2^{15}} \cdot \frac{48}{10000} = \frac{1}{2^{16}} \tag{7.79}$$

Therefore, the error in fixed-point arithmetic is computed as:

$$error[div] = v - \frac{Ri}{2^{16}} \tag{7.80}$$

where the division by 2^{16} is equivalent with a 16 bits right shift.

Example 7.6. *Implement in real-time using 16 bit fixed point arithmetic the following transfer functions:*

- *low-pass filter* $LPF(s) = 100/(s + 100)$;
- *high-pass filter* $HPF(s) = s/(s + 5)$;
- *second-order controller* $K(s) = 22.2(s + 1)(s + 2)/[s(s + 6.66)]$.

assuming that the sampling time of the system is: $T_s = 0.001$ *seconds. Compare the accuracy of the continuous and the discrete forms using numerical simulation.*

Chapter 8

Embedded Safety Loop Development

'From what I have told you, you can deduce what is to come.' — Confucius (551-479 BC)

Summary:

This section is dedicated to embedded safety loop development, used to assure the safety in different industrial applications. The following issues are discussed in this chapter:

- risk assessment and safety levels
- safety function and safety loop
- classification of faults and calculation of probability of failure on demand
- software testing and validation
- memory testing.

8.1 Risk Assessment and Safety Levels

In the past chapters we have focused on embedded control system design and from control viewpoint we required robust stability. This is essential in design but from safety point of view further steps are required.

We can ask questions like:

- what happens if the parameters of our system undergo larger variations than we considered during design?
- what happens if our embedded hardware fails?
- what happens if the supply voltage undergoes large variations or what happens in case of power failure?
- what happens if our embedded software fails?

All these failures might create harm, injury and financial losses. Therefore, the obvious question is: how these failures can be avoided or at least their impact minimized? In this chapter we will try to emphasize good design practices, which - in order to achieve high safety and integrity level - are required.

Every year, millions of people are injured at work, or have their health seriously harmed at the workplace. That means there is a risk at the workplace to be injured or harmed. Within engineering disciplines we define the risk as a product:

Risk = Probability of an accident x consequences (severity) of the accident

The main goal is to reduce the risk at least to the tolerable risk, which is accepted by the society. All work activities are susceptible to risk. That is why risk assessment is so important for a healthy workplace. Risk assessment is a dynamic process that allows enterprises and organizations to put in place a proactive policy of managing workplace risks.

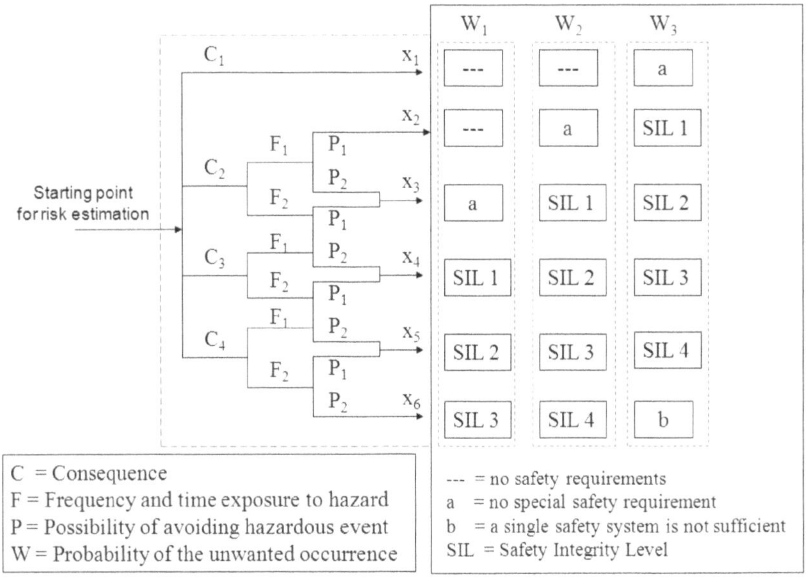

Fig. 8.1 Risk assessment graph.

Risk assessment usually is a stepwise approach as:

- identifying the danger and those at risk
- evaluating the risk and prioritizing the risk
- reduction of the risk, by deciding preventing actions
- taking actions (technical and organizational measures)
- monitoring and reviewing.

The relative level of risk reduction or the target level of risk reduction (introduced in chapter 2) is achieved by the safety loop, which has to achieve a specified safety integrity level (SIL), where SIL 1 is the lowest and SIL 4 is the highest level. Different industry sectors have different SIL levels, although SIL levels are specified in standards there is a very poor harmonization among them. Therefore, in this chapter we will only refer to the IEC 61508 standard, which is a generic standard.

A general risk assessment graph is shown in Fig. 8.1, where depending on the risk parameters, the required safety integrity level is specified. Furthermore, in Table 8.1 the classification of each risk parameter is described.

Table 8.1 Classification of risk parameters

Risk parameter		Classification
Consequence (C)	C_1	Minor injury.
	C_2	Serious permanent injury to one or more persons, death to one person.
	C_3	Death of several persons.
	C_4	Catastrophic consequences, multiple deaths.
Frequency and exposure time to the hazardous zone (F)	F_1	Rare to more often exposure.
	F_2	Frequent to permanent exposure.
Possibility of avoiding hazard (P)	P_1	Possible under certain conditions.
	P_2	Almost impossible.
Probability of the unwanted occurrence (W)	W_1	Very low probability, only a few unwanted occurrences are likely.
	W_2	Low probability, a few unwanted occurrences are likely.
	W_3	Relatively high probability, unwanted occurrences are likely.

8.1.1 Safety Function and Safety Loop

The safety system is said to be functionally safe if random, systematic and common cause failures do not lead to malfunctioning of the safety system and do not result in injury or human death, environment pollution, equipment or production loss. The safety function has to be guaranteed both under normal conditions and in presence of faults.

Defining correctly and simply the safety function is essential in any application, and it requires the definition of the safe state and the temporal constraints to reach the safe state.

Basically, the safety function is realized by a so-called safety loop, which is recommended to be independent form the control loop (see Fig. 8.2). The idea behind is, if the control loop fails then the safety function is triggered and the safety loop shall be able to bring the system into a safe state.

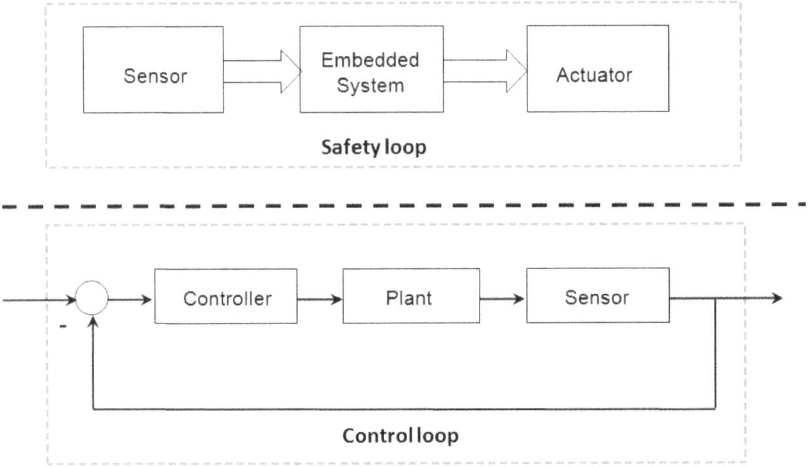

Fig. 8.2 Safety loop and control loop.

The structure of an embedded safety loop is shown in the upper part of Fig. 8.2. It has three main components: sensor, the embedded safety system and the actuator. In other words, when the safety function is triggered the equipment is switched into a safe state or is switched off. This is illustrated in Fig. 8.3, when the emergency button is pushed based on the logic implemented the equipment/machine is switched into a safe state or is switched off.

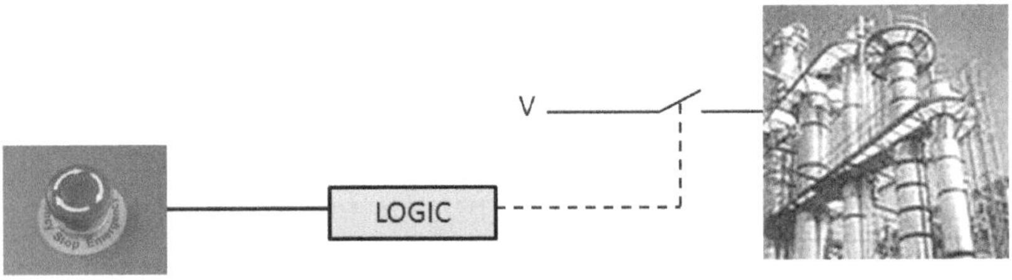

Fig. 8.3 Safety loop.

In many industrial applications an intelligent safety loop is desired, which can differentiate among severity of faults and emergency levels. For example:

- if a minor fault occurs, a warning is issued and the process can still run for a limited time
- if a major fault occurs, a warning is issued and the process switches to a safe mode and runs for a limited time
- if the safety function is triggered or the emergency button is pushed the process switches to a safe state or is switched off immediately.

8.2 Classification of Faults

8.2.1 Fault Categories and Failure Rates

Faults in technological systems are events that happen rarely, often at unexpected moments of time. In Isermann and Balle [45] fault is defined as: an unpermitted deviation of at least one characteristic property or parameter of the system from the acceptable/usual/standard condition.

If faults could timely be detected and diagnosed in many cases it is possible to subsequently reconfigure the system so that it can safely continue its operation, until the time when it can be switched off for maintenance.

Faults can be classified according to their location of occurrence, such as: sensor faults, component faults and actuator faults. Typical share of failure rates experienced by industrial plants - according to TUV Rheinland (2008) - is shown in Fig. 8.4 [85].

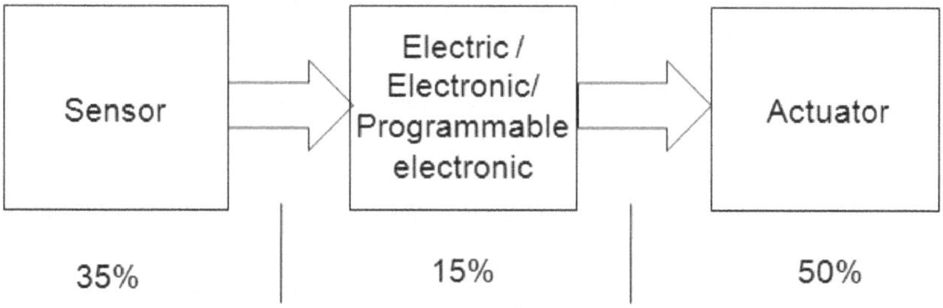

Fig. 8.4 Failure rates in industrial systems.

According to their cause, failures can be divided in two categories:

- systematic failures. These failures can be eliminated only by alteration of design, production process, operation mode or operation instructions. Software faults fall into this category.
- random failures. These failures occur at random time and the cause cannot be defined exactly. Hardware failures belong to this category.

In Fig. 8.5 the cause of systematic and human failures are show (source U.K. Health and Safety Executive), without identifying the technology sector.

Random failures are investigated using statistical methods and a so-called Mean Time To Failure (MTTF) is defined, which is measured in hours [h]. Furthermore, the component failure rate λ is defined as inverse of Mean Time To Failure (MTTF) and is measured in failure in time [fit]. Where $1[fit] = [10^{-9} failure/hour]$.

$$\lambda = \frac{1}{MTTF} \tag{8.1}$$

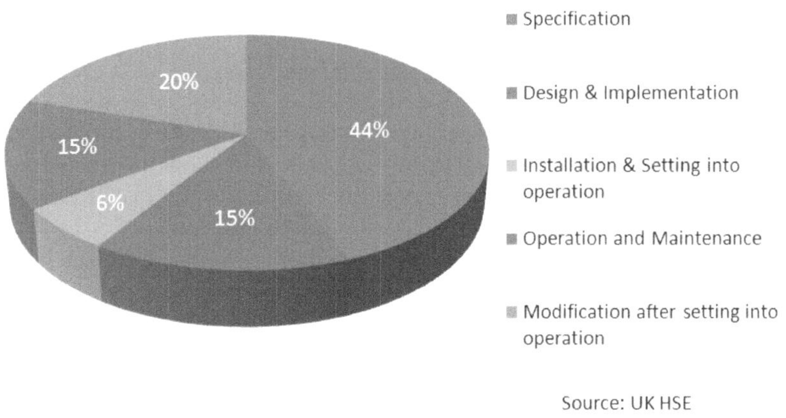

Fig. 8.5 Cause of systematic failure.

The failure rate varies in time as shown in Fig. 8.6. Three distinct phases can be identified as: early failure phase, useful lifetime and end of lifetime.

During the useful lifetime the failure rate is approximately constant. Such values are defined in the SN 29500 (Siemens Norm), which specifies the failure rates for different components at reference conditions. If there are deviations from the reference condition (see different operating condition) the failure rate shall be recalculated. The operating temperature usually has a big impact on the failure rate.

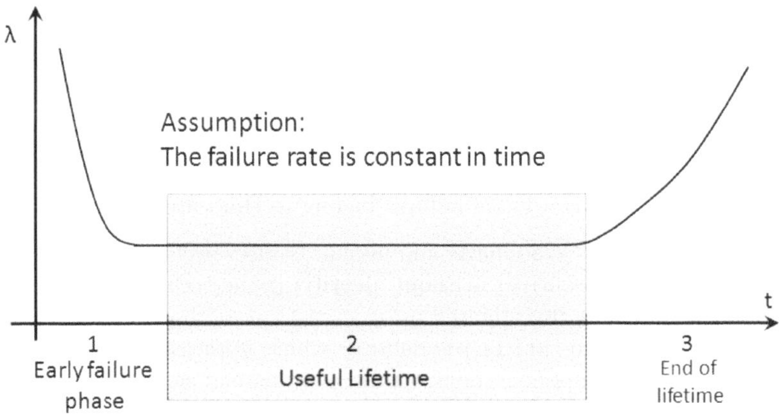

Fig. 8.6 Time variation of the failure rate.

We remark, that the failure rates λ, which will be used in calculations are constant values, which means - for simplicity - we neglect the time variation of the failure rate.

The probability of failure on demand at time moment t is expressed as:

$$PFD(t) = 1 - e^{-\lambda t} \approx \lambda t \tag{8.2}$$

the approximation is valid if $\lambda t << 1$, condition which is fulfilled in most of the cases in practice.

Furthermore, is possible to calculate the average probability of failure on demand over a period of time, i.e. proof test interval T_p:

$$PFD_{avg}(T_p) = \frac{1}{T_p} \int_0^{T_p} 1 - e^{-\lambda t} dt = \frac{1}{2} \lambda T_p \tag{8.3}$$

We can also calculate the probability of failure per hour (PFH) as a ratio of the average probability of failure on demand and average period of time, which in our case is $T_p/2$:

$$PFH \approx \frac{PFD(T_p)}{T_p} = \frac{PFD_{avg}(T_p)}{T_p/2} \tag{8.4}$$

8.2.2 Safe Failure Fraction, Diagnostic Coverage and Hardware Fault-Tolerance

Safe Failure Fraction

From safety point of view failures are divided as (see Fig. 8.7):

- safe failure (denoted by λ_S)
- dangerous failures (denoted by λ_D), which are furthermore divided into dangerous detectable (denoted by λ_{DD}) and dangerous undetectable failures (denoted by λ_{DU}).

Having the failure rate defined for the components, we would like to calculate the so-called safe failure fraction (SFF) for the safety loop.

The safe failure fraction is defined as a ratio:

$$SFF = \frac{\lambda_S + \lambda_{DD}}{\lambda_{TOT}} \tag{8.5}$$

where: λ_S is the sum of safe faults, λ_{DD} is the sum of dangerous and detected faults and λ_{TOT} is the sum of all faults. As a remark, only components which are part of the safety function (loop) are included.

For simple systems, built form for simple components the failure mode can be easily defined and failure mode effective analysis can be applied to calculate the safe failure fraction. However for systems with complex components (such as microcontrollers, memory, etc.) failure modes are difficult to define. In such a case, according to IEC 61508-6 Annex C: for complex components, where a detailed analysis of each failure is not possible, a division of failures into 50% safe and 50% dangerous is generally accepted.

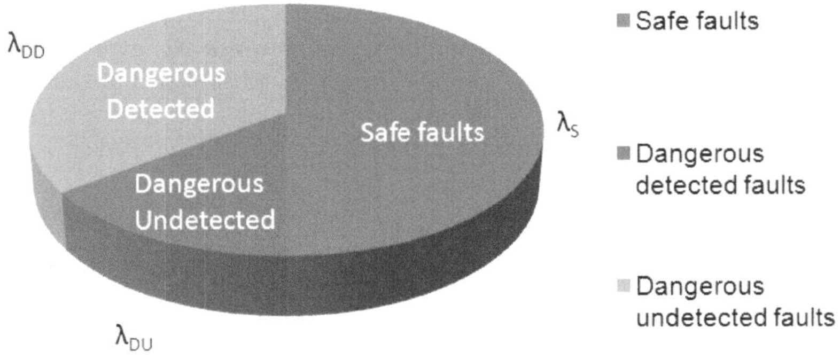

Fig. 8.7 How failures are divided.

Diagnostic Coverage

Obviously, some parts of dangerous failures is possible to detect using different on-line diagnostic methods. Therefore, the diagnostic coverage denoted by DC is defined as:

$$DC = \frac{\lambda_{DD}}{\lambda_{DD} + \lambda_{DU}} \tag{8.6}$$

Different diagnostic methods offers different diagnostic coverage, however in practice and in standards for simplification diagnostic coverage is divided into four basic ranges:

- $DC < 60\%$ means no diagnostic coverage
- $60\% \leq DC < 90\%$ means low diagnostic coverage
- $90\% \leq DC < 99\%$ means medium diagnostic coverage
- $DC \geq 99\%$ means high diagnostic coverage.

Example 8.1. *Let us assume that a ROM memory device has a failure rate of 100 [fit]. Furthermore, let us assume that on-line memory diagnostic is performed, which assures medium diagnostic coverage. Calculate the safe failure fraction (SFF) for the ROM memory.*

The ROM memory device belongs to complex components, according to IEC61508 (see also previous section) the failures are divided into 50% safe and 50% dangerous failures thus we have:

$$\begin{aligned} \lambda_{TOT} &= 100[fit] \\ \lambda_S = \lambda_{TOT}/2 &= 50[fit] \\ \lambda_D = \lambda_{TOT}/2 &= 50[fit] \end{aligned} \tag{8.7}$$

Since on-line diagnostic is performed with medium diagnostic coverage, we have $DC = 90\%$. Therefore, we obtain (see Fig. 8.8):

$$\begin{aligned} \lambda_{DD} &= DC * \lambda_D = 45[fit] \\ \lambda_{DU} &= (1 - DC) * \lambda_D = 5[fit] \end{aligned} \tag{8.8}$$

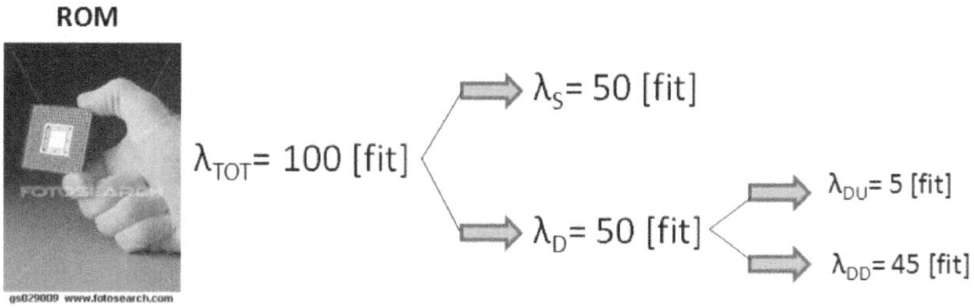

Fig. 8.8 Division of ROM failure rate.

Then we can calculate the safe failure fraction (SFF):

$$SFF = \frac{\lambda_S + \lambda_{DD}}{\lambda_{TOT}} = \frac{50 + 45}{100} = 95\% \tag{8.9}$$

Hardware Fault-Tolerance

For a given system, based on the components failure rates and considering the implemented diagnostic measures the safe failure fraction (SFF) can be computed. However, in industrial applications usually the required safety and integrity level (SIL) and the safe failure fraction (SFF) of the safety loop is specified in standards. The safety and integrity level, which can be achieved is also dependent of the hardware architecture.

Definition 8.1. *The hardware is fault-tolerant (HFT) of order N if N+1faults can cause the loss of safety function. Therefore, a single channel system has* $HFT = 0$*, and a dual-channel system has a* $HFT = 1$*.*

Table 8.2 shows SIL levels for low-complexity systems defining also architectural requirements for the safety loop, based on hardware fault-tolerance.

In a similar way, Table 8.3 shows the SIL level for high-complexity systems. Safety loops containing embedded systems usually fall into this category. In practice, is said that if at least one system component is high-complexity the system falls into the category of high-complexity systems.

The Table 8.3 also suggests that in order to achieve SIL3 level for high-complexity systems in most of the cases at least a dual-channel architecture is required.

Table 8.2 SIL levels for low-complexity systems

	Hardware Fault-Tolerance		
Safe Failure Fraction (SFF)	0	1	2
less than 60%	SIL 1	SIL 2	SIL3
$\geq$ 60%	SIL 2	SIL 3	SIL 4
$\geq$ 90%	SIL 3	SIL 4	SIL 4
$\geq$ 99%	SIL 4	SIL 4	SIL 4

Table 8.3 SIL levels for high-complexity systems

	Hardware Fault-Tolerance		
Safe Failure Fraction (SFF)	0	1	2
less than 60%	Not allowed	SIL 1	SIL2
$\geq$ 60%	SIL 1	SIL 2	SIL 3
$\geq$ 90%	SIL 2	SIL 3	SIL 4
$\geq$ 99%	SIL 3	SIL 4	SIL 4

8.3 Calculation of Probability of Failure on Demand

In this section we are going to show how to calculate the probability of failure on demand (PFD) and probability of failure per hour (PFH), which are important characteristics of safety systems. The probability of failure calculation depends on the operation mode of the safety related system.

The **low demand mode** is defined as the operation mode where the frequency of the operation of the safety related system in not greater than once per year and not greater than twice the proof test interval denoted by T_p. The proof test interval is defined as a periodic test performed to detect dangerous failures in a safety related system.

8.3.1 PFH Calculation for Single-Channel System

In case of single-channel systems, where the safety function is operating in low demand the average probability of failure on demand is calculated as shown below.

Single-channel system without diagnostic - low demand

$$PFD_{avg}(T_p) = \frac{1}{2}\lambda_D T_p \tag{8.10}$$

Single-channel system with diagnostic - low demand

$$PFD_{avg}(T_p) = \frac{1}{2}\lambda_{DU} T_p \tag{8.11}$$

In case of systems, where the safety function is operating in **high demand mode** the probability of failure on hour (PFH) is calculated.

Single-channel system without diagnostic - high demand

$$PFH = \lambda_D \tag{8.12}$$

Single-channel system with diagnostic - high demand

$$PFH = \lambda_{DU} \tag{8.13}$$

Furthermore, the probability of failure on demand can be calculated, where T_p is the proof test interval. After the proof test the system can be considered as new.

For each component, which is part of the safety loop the PFH shall be calculated and in case of a single channel system (see Fig. 8.9) the sum of these values gives the PFH of the whole safety loop.

$$PFH = \lambda_{D\_Sensor} + \lambda_{D\_Logic} + \lambda_{D\_Actuator} \tag{8.14}$$

where

$$\lambda_{D\_Logic} = \lambda_{D\_Inp\_Cir} + \lambda_{D\_Pow\_Sup} + \lambda_{D\_Micro} + \lambda_{D\_Mem} + \lambda_{D\_Out\_Rel} \tag{8.15}$$

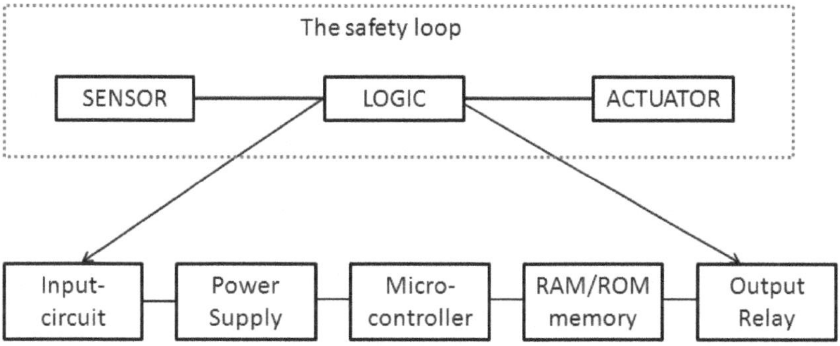

Fig. 8.9 Single-channel safety loop.

8.3.2 PFH Calculation for Dual-Channel System

First, let us consider a dual-channel system (shown in Fig. 8.10) without diagnostic and without fault detection implemented in the single channels.

Example 8.2. *For a dual-channel system operating at high demand, let assume that PFH for channel one is:* $PFH_1 = 5 \cdot 10^{-4}[1/h]$ *and for channel two is same:* $PFH_2 = 5 \cdot 10^{-4}[1/h]$. *Calculate PFH for the dual-channel system.*

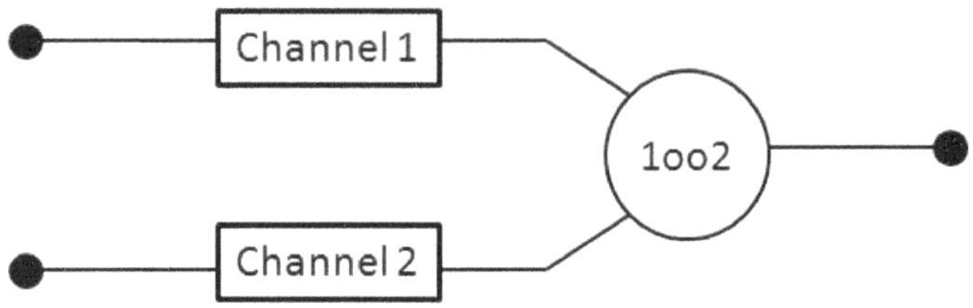

Fig. 8.10 Dual-channel 1oo2 block diagram.

It would be expected that the probability of failure for a dual-channel system would be:

$$PFD(t) = (PFH_1 \cdot t) \cdot (PFH_2 \cdot t) \tag{8.16}$$

with PFH calculated for $t = 1[h]$ as:

$$PFH = \frac{PFD(t)}{t} = 25 \cdot 10^{-8}[1/h] \tag{8.17}$$

However, practice - based on collected experimental failure rates - shows different values. The primary reason for the difference between the theoretical result and the experimental one, is that channel failures are not totally independent and Common Cause (CC) failures have a significant influence on the probability of the total system failure (see 8.11).

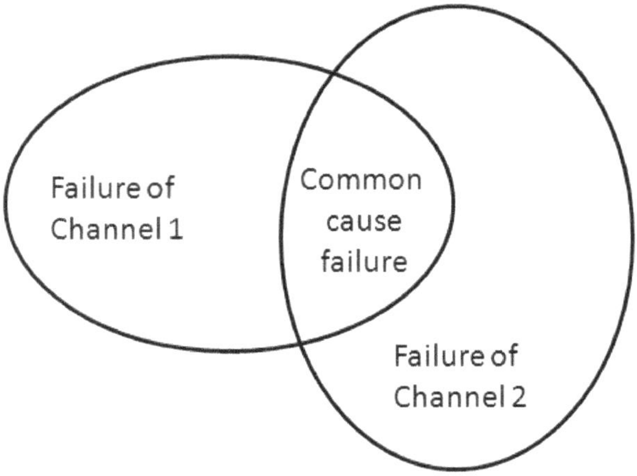

Fig. 8.11 Common cause failure.

Equation 8.16 is valid if the two channels are totally independent of each other. For example, the probability of failure in channel one is independent of the probability of failure in channel two, that means the probability of common cause is zero.

Common cause failures are due to a single cause and affect more than one channel (in our case both channels). In other words due to one cause a failure in one channel is followed by the same failure in the other channel, either at the same time or some time later.

It might be due to:

- external stress such as excessive temperature, high electromagnetic interferences, etc.
- systematic design failures such as high complexity of the design or missing experience with the new technology.

Continuing the example above let us assume that 4% of the failure rate in each channel is due to the common cause. In this case the probability of failure in one channel can be written as a sum of probability of failure due to random cause - in this example 96% - and probability of failure due to the common cause - in this example 4%.

Thus we have:

$$PFD_1(t) = 0.96 \cdot PFD_1(t) + 0.04 \cdot PFD_1(t) \tag{8.18}$$

Now we might calculate the probability of failure rate of the dual-channel system:

$$PFD(t) = (0.96 \cdot PFD_1(t))^2 + 0.04 \cdot PFD_1(t) \tag{8.19}$$

with PFH calculated for $t = 1[h]$ as:

$$PFH = \frac{PFD(t)}{t} = (4.8 \cdot 10^{-4})^2[1/h] + 0.2 \cdot 10^{-4}[1/h] \tag{8.20}$$

We can remark, that the common cause failure plays a major part in the PFH, therefore in dual-channel (or multiple-channel) systems cannot be neglected.

The common cause failure shall be considered for each main part of the safety loop: sensors, logic and actuators as shown in Fig. 8.12.

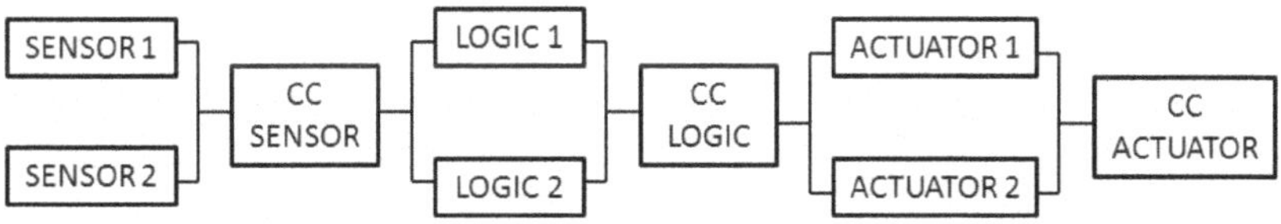

Fig. 8.12 PFH dual-channel system.

Common cause failures can be reduced significantly by separation (segregation) of channels, or by using so-called diverse redundancy, which is more effective than homogeneous redundancy but is more expensive. Systematic design measures such as competence, training, environmental testing, etc. can also reduce the rate of common cause failure.

The common cause failure in the IEC61508 standard is quantified in a so-called β- factor. The standard shows how this β-factor is calculated based on diagnostic coverage and diagnostic test interval for programmable electronics as well as sensors and final elements. The β-factor represents a fraction of the failure rate, which are due to the common cause.

Therefore, we can write that the variation in time of the probability of failure on demand is:

$$PFD(t) = [(1-\beta)\lambda_D \cdot t]^2 + \beta\lambda_D \cdot t \tag{8.21}$$

If the dangerous failures rates in channel 1 and channel 2 are different and are denoted by λ_{D1} and λ_{D2}, we might write:

$$PFD(t) = [(1-\beta)\lambda_{D1} \cdot t] \cdot [(1-\beta)\lambda_{D2} \cdot t] + \beta\left(\frac{\lambda_{D1}+\lambda_{D2}}{2}\right) \cdot t \tag{8.22}$$

The average value of the probability of failure on demand is calculated as:

$$PFD_{avg}(T_p) = \frac{1}{T_p}\int_0^{T_p} PFD(t)dt \tag{8.23}$$

where T_p is the proof test interval.

The probability of failure per hour is calculated as:

$$PFH = \frac{PFD_{avg}(T_p)}{T_p/2} \tag{8.24}$$

Dual-channel system with diagnostics

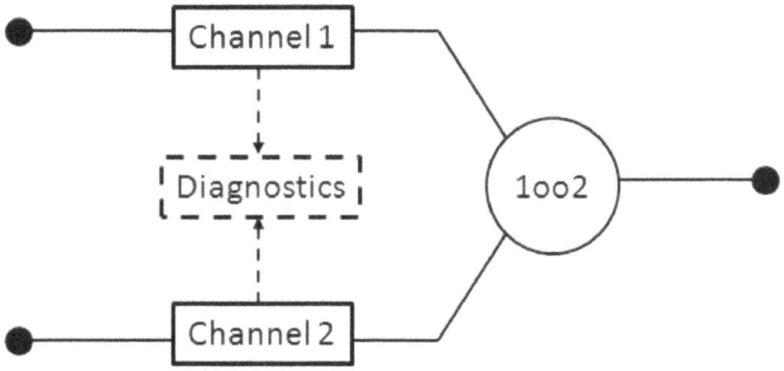

Fig. 8.13 Dual-channel 1oo2 with diagnostics block diagram.

In case of dual-channel systems with diagnostics (shown in Fig. 8.13) the IEC61508 standard -related to the common cause - distinguishes between β-factor for dangerous undetected failures and β_D-factor for dangerous detected failures.

In this case, in addition to the dangerous undetectable failures simultaneous dangerous detectable failures as well as their common cause can lead to the loss of safety function. Therefore, the variation in time of the probability of failure on demand is:

$$PFD(t) = PFD_{DU}(t) + PFD_{DD}(t) \tag{8.25}$$

where

$$PFD_{DU}(t) = [(1-\beta)\lambda_{DU1}\cdot t]\cdot[(1-\beta)\lambda_{DU2}\cdot t]+\beta\left(\frac{\lambda_{DU1}+\lambda_{DU2}}{2}\right)\cdot t \tag{8.26}$$

$$PFD_{DD}(t) = [(1-\beta_D)\lambda_{DD1}\cdot t]\cdot[(1-\beta_D)\lambda_{DD2}\cdot t]+\beta_D\left(\frac{\lambda_{DD1}+\lambda_{DD2}}{2}\right)\cdot t \tag{8.27}$$

The average value of the probability of failure on demand is calculated as:

$$PFD_{avg} = \frac{1}{T_p}\int_0^{T_p} PFD_{DU}(t)dt + \frac{1}{T_D}\int_0^{T_D} PFD_{DD}(t)dt = PFD_{avg}(T_p) + PFD_{avg}(T_D) \tag{8.28}$$

where T_p is a proof test interval (i.e. 1 year) and T_D is the diagnostics test interval (i.e. 1 hour) in both channels.

The probability functions are integrated over different time ranges, since is assumed that the dangerous detectable failures will be found during the diagnostics test interval.

If we assume that $\lambda_{DU1} = \lambda_{DU2} = \lambda_{DU}$ and $\lambda_{DD1} = \lambda_{DD2} = \lambda_{DD}$ we obtain:

$$PFD_{avg}(T_p) = \frac{1}{3}(1-\beta)^2\lambda_{DU}^2 T_p^2 + \frac{1}{2}\beta\lambda_{DU}T_p \tag{8.29}$$

$$PFD_{avg}(T_D) = \frac{1}{3}(1-\beta_D)^2\lambda_{DD}^2 T_D^2 + \frac{1}{2}\beta_D\lambda_{DD}T_D \tag{8.30}$$

The probability of failure per hour is calculated as:

$$PFH = \frac{PFD_{avg}(T_p)}{T_p/2} + \frac{PFD_{avg}(T_D)}{T_d/2} \tag{8.31}$$

For PFH calculation the IEC 61508 standard considers a similar approach. The state transition diagram is shown in Fig 8.14, where the states are defined as:

- State 0: the safety system is operating correctly, or a safe failure has occurred
- State 1: a dangerous failure that cannot be detected by diagnostic tests has occurred
- State 2: a dangerous failure that can be detected by the diagnostic tests has occurred.

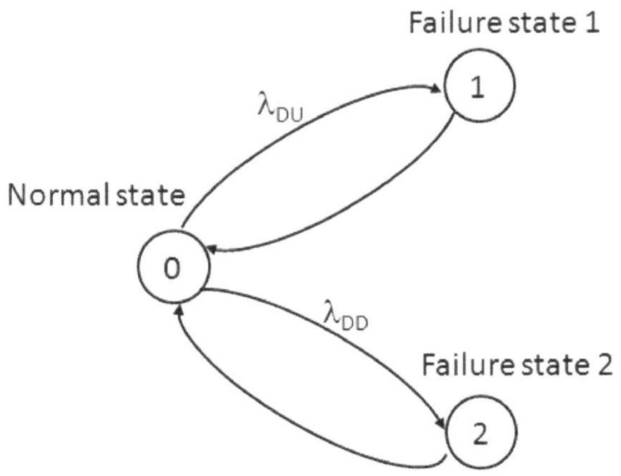

Fig. 8.14 State transition diagram.

If the safety system is in safe state it can operate the safety function, but when the safety system is in state 1 or state 2, the safety system is in failed state and the safety function cannot be operated. The channel equivalent mean down time t_{CE} is the expected time the system is in failure state 1 or failure state 2.

The formulas for PFH calculation given in the standard are different from the formulas presented here, but the results are very close to each other. Furthermore, in the standard a procedure is presented how β and β_D factors are derived. For further details, the interested reader shall consult reference [84].

There are situations when the sensors are connected to the embedded safety system via a safety communication network. One such safety communication protocol is the CANopen Safety, which main features are summarized below:

CANopen Safety Communication

The CANopen Safety is based on the CANopen communication protocol. The idea of safety in CAN communication is not to ensure, absolutely no errors and faults. The goal is to detect all possible errors and react in a predictable (safe) way.

In CANopen Safety communication networks the safety-relevant data transfer is performed by means of Safety Related Data Objects (SRDO). An SRDO consists in two CAN data frames with identifiers, which shall be different in at least two bit positions. The user data in both transmissions is redundant, i.e. the meaning of the data is same, but the data on the second transmission is inverted bitwise as shown in Fig. 8.15.

The SRDO shall be only valid, if both CAN frames are received properly (without failure and in time). The SRDO are refreshed periodically within

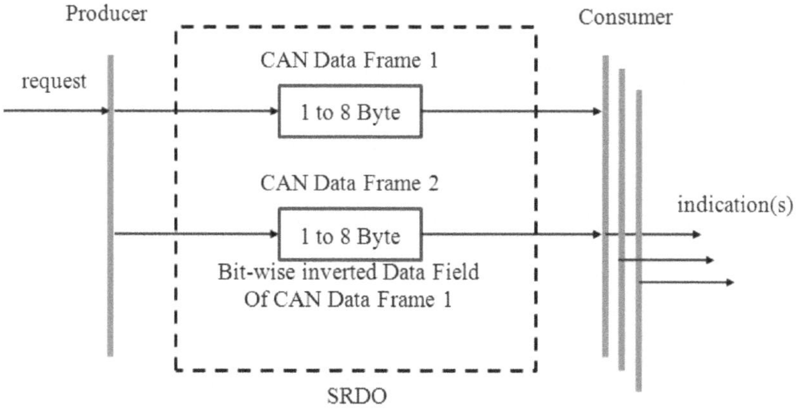

Fig. 8.15 CANopen Safety protocol - Safety Relevant Data Object.

the refresh time denoted by SCT. If the SCT expires, the embedded safety controller shall bring the system in the safe state.

Furthermore, both CAN frames that make an SRDO shall be received correctly within the safety re-transmit time denoted by $SRVT$. Normally both CAN frames are transmitted with minimum delay. If the second CAN frame is not being received within $SRVT$, the network may have reduced transmission capabilities. If the $SRVT$ expires, the embedded safety system shall bring the system into a safe state.

The safety refresh time and safety re-transmit time is shown in Fig. 8.16.

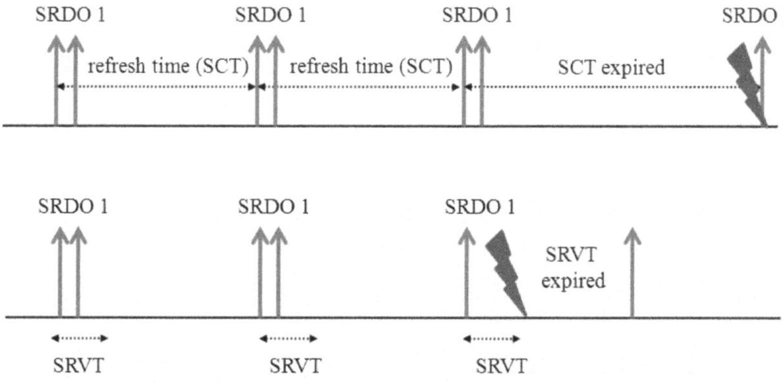

Fig. 8.16 Safety Relevant Data Object (SRDO) Timing.

8.3.3 Operating Demand of the Safety Function

According to IEC61508 the safety and integrity level is defined function on the operating demand of the safety function.

In case of the low demand operation the SIL levels are summarized in Table 8.4.

Table 8.4 Low demand mode of operation

Safety and integrity level (SIL)	Average probability of failure on demand (PFD_{avg})
SIL 4	$10^{-5} < PFD_{avg} \leq 10^{-4}$
SIL 3	$10^{-4} < PFD_{avg} \leq 10^{-3}$
SIL 2	$10^{-3} < PFD_{avg} \leq 10^{-2}$
SIL 1	$10^{-2} < PFD_{avg} \leq 10^{-1}$

In case of the high demand or continuous mode of operation the SIL levels are summarized in Table 8.5. We remark that most of the industrial safety related functions have high demand or continuous operation mode.

Table 8.5 High demand mode of operation

Safety and integrity level (SIL)	Probability of failure per hour (PFH)
SIL 4	$10^{-9} < PFH \leq 10^{-8}$
SIL 3	$10^{-8} < PFH \leq 10^{-7}$
SIL 2	$10^{-7} < PFH \leq 10^{-6}$
SIL 1	$10^{-6} < PFH \leq 10^{-5}$

The difference of 10^4 in the values in the two tables is due to the fact that the standard defines low-demand as "no greater than one demand per year" and one year is approximated to 10^4 hours.

In case of high demand operation, in addition to the PFH the average probability of failure on demand PFD_{avg} is also calculated.

As conclusion, for a given architecture, knowing the components failure rate and the implemented diagnostics is possible to calculate the safe failure fraction (SFF), probability of failure per hour (PFH), average probability of failure on demand PFD_{avg}. Based on these values is possible to verify if the required SIL level is fulfilled or not.

However, having fulfilled the SIL level, does not mean that our safety system is ready to be deployed.

The calculation above considered only the random faults, in the next section we will discuss how systematic faults - especially software faults - can be minimized or avoided.

8.4 Software Testing and Validation

Software testing plays an important role in the software development. Early detection of the software errors can save costs and reduce significantly development time. Since software development in most cases is an iterative process (requirements can change or new requirements might be defined) it is highly recommended to apply test automation. For every new software version the previous tests have to be re-run (this is called regression tests) to assure that the new changes have not introduced errors. Test automation has several advantages such as:

- more testing can be completed in less time
- large number of test cases can be executed without human intervention
- tests cases are valuable - once they are created they can be reused.

Testing methods are grouped into different categories based on different criterion and are mentioned below.

8.4.1 Static and Dynamic Software Testing Methods

Software testing methods - depending if the software under test is executed or not - can be divided into:

- static testing methods
- dynamic testing methods.

Static testing is a testing method where the software code is inspected but the code is not run. It is mainly a syntax checking of the code and/or manually reviewing the code or document to find errors [34].

Code reviews, code inspection and walk-through fall into the category of static testing. Formal code inspections are probably the most important tool to reduce quickly software bugs. The code inspection is based on the well-known fact that "two heads are better than one". The main goal is to identify and remove software bugs before testing.

For example:

- IBM found inspection gave 23% increase in productivity and 38% reduction in bugs detected after unit test [34]
- HP found that 80% of the errors detected during inspection were unlikely to be caught by testing [34].

Remark 8.1. *Although, code inspection may cost up to 20% more time, debugging can shrink by an order of magnitude. There is no known better way to find bugs than through code inspections! Skipping the code inspection is a sure sign of amateurism [34].*

Dynamic testing is a software testing method where the software under tests is complied and run. Dynamic testing means working with the software: giving input values and checking if the output is as expected. Unit tests, integration tests, system tests and acceptance tests are few of the dynamic testing methodologies.

8.4.2 Software Development Stage Related Testing Methods

Depending at which software development stage is executed we define: unit tests, integration tests and system tests and acceptance tests.

Unit tests are performed for each individual module of the source code and is verified if they are working properly.

Integration tests are performed to check the connectivity among the individual modules and test the functionality.

There are different integration approaches as mentioned below:

- big-bang integration testing (all individual modules are integrated together in one step)
- top-down integration. High level modules are integrated and tested first, for example testing from main module to sub-module. If a module or sub-module is not ready for integration stubs are used as temporary module.
- bottom-up integration testing. Low level modules are integrated and tested first, i.e. from sub-module to main module. If a sub-module is not available a temporary module (driver) is used.

The big-bang integration approach has several disadvantages such as: integration is performed at a late stage, errors are more difficult to locate since all modules were integrated at once.

System tests are conducted on a complete system in order to evaluate the systems compliance with the specified requirements. System testing falls in the category of black box testing since does not require knowledge of the inner design or software code.

In case of embedded systems, Hardware-in-the-Loop (HIL) tests can be used in a pre-phase of system tests. Since HIL testing is essential for embedded systems, it will be discussed later in a special section.

Acceptance tests often part of the contractual agreement - are tests run by the end user in their own environment to evaluate whether the system meets their criterion. Based on these tests the customer accepts or rejects the software.

8.4.3 Black-Box and White Box Testing

Testing methodologies - depending if the internal structure of the software code is accessible or not - are divided into two large categories:

- black-box testing
- white box testing.

Black-box testing is a software testing method, which does not require the specific knowledge of the internal structure of the software code. The internal structure of the test object in not known. Test cases are derived based on defined requirements and specifications. The test designer selects valid and invalid inputs and determines the correct output. These tests can be done at all software testing levels: unit, integration, system and acceptance level. In most of the cases is a functional test so is mainly used at higher testing levels, but is also done at unit testing level.

White-box testing is a software testing method, which requires the specific knowledge of the internal structure of the software code. In white-box testing an internal perspective of the system, as well as programming skills are required to design test cases. The software tester chooses inputs to exercise paths through the code and determines the correct output. In most of the cases this is done at unit level, but is possible to perform it at integration and system level, too. It is a very effective method to detect software errors, but it might not detect missing requirements or unimplemented parts of the specifications.

8.4.4 Software Testing Metrics – Code Coverage

Software testing is a time consuming and costly process. In the aim to make software testing effective, test strategies shall be defined and should be used to derive test cases (suites). Software testing metrics such as code coverage, shall be used to analyze the adequacy of the defined test suites. Coverage reports can point out easily inadequate test suites, but 100% test coverage cannot guarantee error-free software.

The code coverage model defines the parts of an implementation that shall be exercised during the tests. The coverage is a metric (expressed in percentage), defined as a ratio between the exercised parts and the defined parts, specified by the coverage model.

To derive the testing coverage, the code shall be instrumented, that means it will require the access to internal structure of the software code. Therefore, testing coverage models defined below can be applied only in case of white box testing methods.

Different coverage models have been defined and used in the software industry. However, the most relevant and frequently used are:

- statement coverage, denoted as C0
- decision coverage, denoted as C1
- condition coverage, denoted as C2.

In order to define coverage models, the software code is abstractly represented by a so called control flow graph (CFG), which is a directed graph.

The control flow graph has nodes and edges:

- the node corresponds to a code segment and is labeled by letters or numbers
- the edge corresponds to a conditional transfer between code segments and is represented by an arrow.

Furthermore, the entry point of the graph is represented by the entry node, without inbound edges and the exit point is represented by the exit node, without outbound edges.

The control flow graph is based on the following concepts:

- code segment
- predicate
- condition.

The code segment consists of one or several statements without conditionally executed statements. The last statement in the segment must be a conditional expression, loop control, a break, an exit, etc. Basically, the last part in the segment selects another segment but does not include any subsequent segment code.

The predicate is a conditional expression, which is evaluated true or false. Predicates are used in control statements: if, case, do, while, do until, etc. After the predicates is evaluated the control is transferred to another code segment. A predicate with multiple conditions is called compound predicate. Furthermore, a condition corresponds to each boolean operator in the predicate expression.

Loop structures can be transformed into control flow graphs as shown in Fig. 8.17., Fig. 8.18. and Fig.8.19.

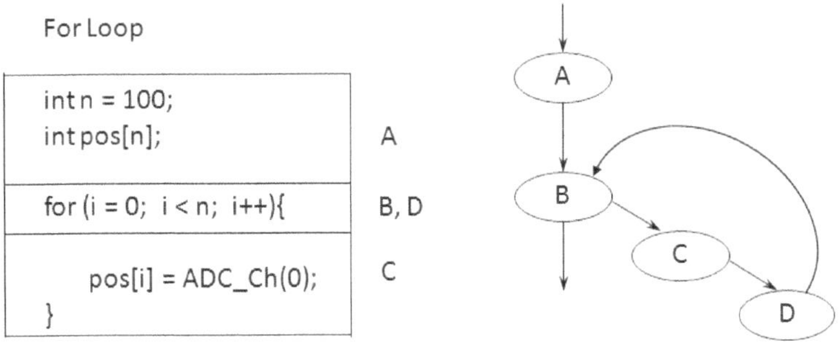

Fig. 8.17 For Loop - control flow graph.

How a control flow graph is built, is illustrated in Fig. 8.20. The left side of the figure shows the program flow using a flow chart representation and the right side shows the corresponding control flow graph. In order to identify

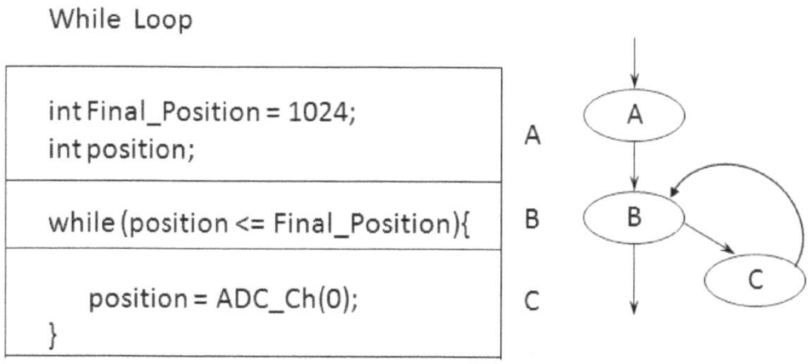

Fig. 8.18 While Loop - control flow graph.

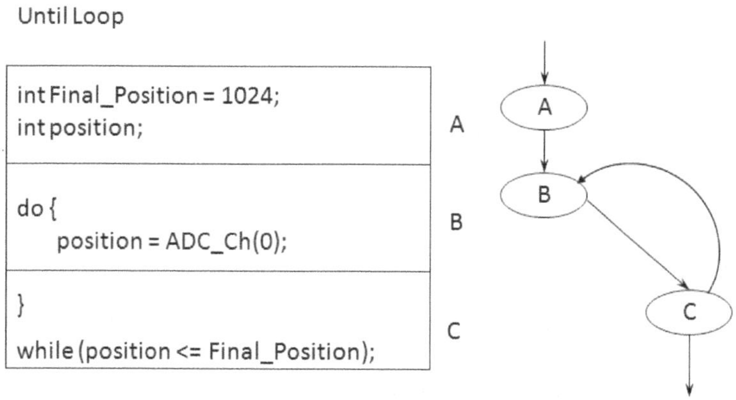

Fig. 8.19 Until Loop - control flow graph.

easily the nodes corresponding to code segments, the two graphs are placed next to each other.

Loops and paths are defined in the graph. Loops are represented by segments within parentheses, followed by an asterisk to show that this group may iterate n times. The path is denoted by the nodes that comprise the path. An entry-exit path is starting with the entry node and is ending with the exit node.

The control flow graph shown in Fig. 8.20 has two entry/exit paths, denoted by:

- ABFG
- AB(CD)*EG

In the light of control flow graph representation:

- 100% statement coverage (C0) means that every statement in the the program is executed at least once or every node in the control flow graph is executed at least once

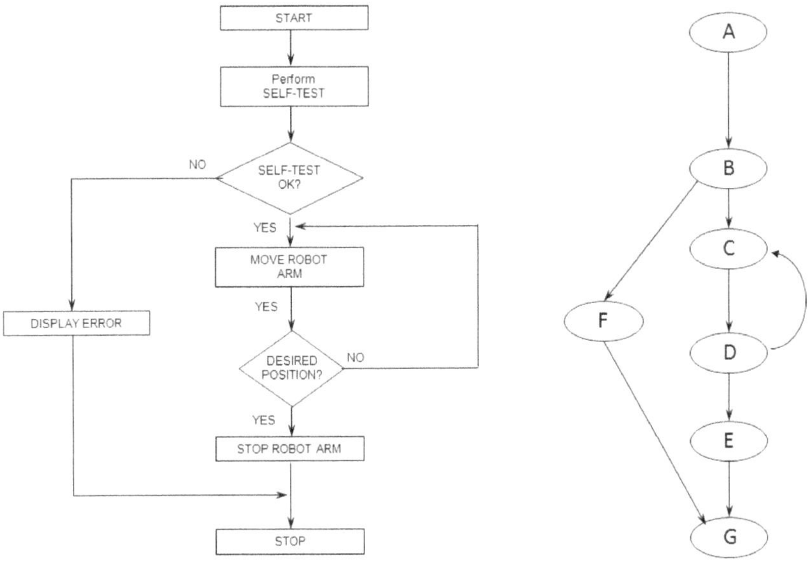

Fig. 8.20 Program flow and control flow graph.

- 100% decision (or branch) coverage (C1) means that every compound predicate (decision) is evaluated true and false at least once or every edge in the control flow graph is executed at least once
- 100% condition coverage (C2) means that each condition in each compound predicate (decision) is evaluated true and false at least once or every edge in the control flow graph is executed at least once.

It also possible to define path coverage (C3), which means that all possible paths in the program are exercised at least once.

As a remark, white box testing has limitations such as: testing of every path and loop in a program is not possible. This means that exhaustive testing is not possible for large systems. However, this does not mean that white box testing is not an effective testing method. By selecting important logical paths and data structures for testing is practically possible and effective.

8.4.5 Hardware in the Loop Tests

Hardware-in-the-loop (HIL) is an effective platform for developing and testing real-time embedded systems. The basic idea is to replace the real system by a real-time plant model, which is running on the hardware (see Fig. 8.21). This is why is called hardware-in-the-loop (HIL). The embedded system under tests is connected to the HIL, which shall behave like the real system. Thus, tests performed using HIL should be very close to tests with the real system.

The main benefits of hardware-in-the-loop testing are:

- improved software quality
- different errors can be injected in real-time
- tests can be repeated easily and tests can be automated
- more effective software development is possible. Design issues are identified earlier, validation time becomes shorter. Lower development costs, lower costs to innovate.

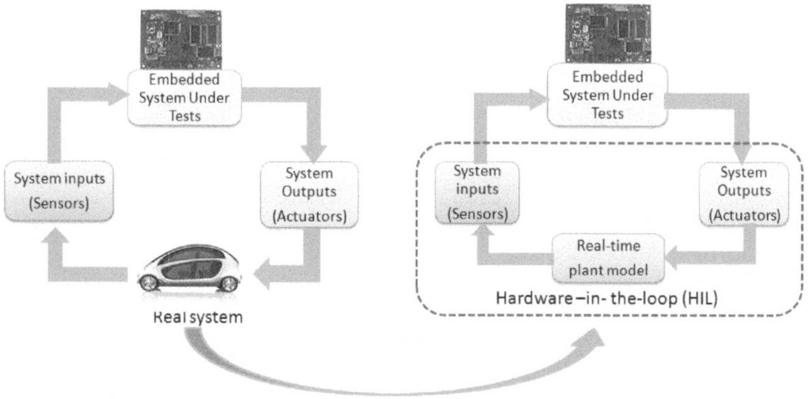

Fig. 8.21 Hardware in the loop (HIL) test concept.

Next, we will try to exemplify the main points of HIL testing via a example.

Example 8.3. *Let us consider the safety system of a linear positioning system shown in Fig. 8.22. Present a possible HIL testing environment for the embedded safety system.*

The positioning belt is driven by motor M. If the overspeed limit is exceeded (see Fig. 8.22) then the safety function is triggered: the motor M is switched off and the safety relays open the safety line and $Brake1$ and $Brake2$ are released. The reaction time of the embedded safety system from the detection of the overspeed condition till opening the safety relays is specified to be 100 ms. This time does not include the brake release time and the time to stop the positioning system.

The overspeed limit for different position values is shown in Fig. 8.22 by the dashed line. In the same figure, the maximum admissible speed for different position values is marked by the solid line.

The speed and position information is provided by a dual-channel encoder. The speed information is represented on 2 bytes and the position information on 4 bytes and are sent to the embedded safety system using CANopen Safety communication protocol.

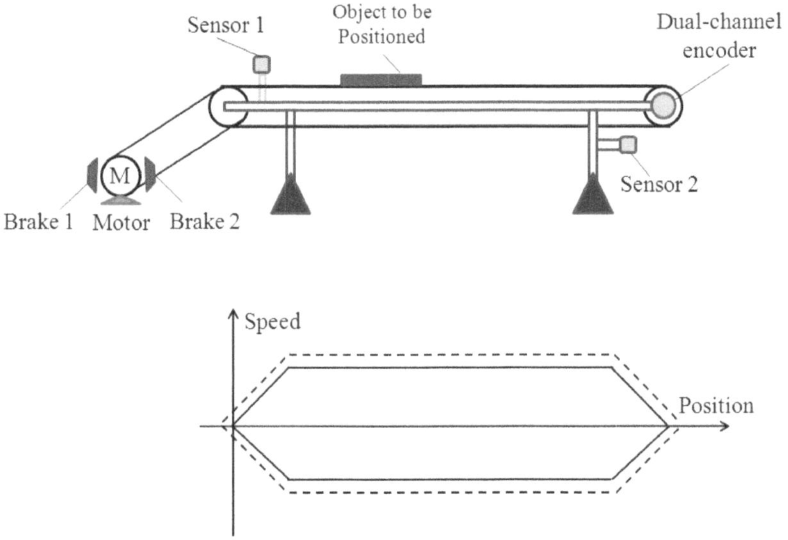

Fig. 8.22 Linear positioning system and its safe operation region.

Two proximity sensors (dual-channel), denoted by *Sensor*1 and *Sensor*2, which provide zero initial position are wired to the embedded safety system.

The embedded safety system contains two safety relays, which open the safety line if the safety function is triggered or an error is detected in the safety loop (sensor + logic + safety relays).

A possible hardware architecture of the safety system is shown in Fig. 8.23, where we remark that in addition to the internal diagnostics, the safety relays are also monitored.

The safety-related embedded system inputs and outputs are shown in Fig. 8.24.

Some of the requirements imposed on the HIL environment might be:

- reproduce the normal operating conditions of the linear positioning system
- create overspeed conditions
- inject different errors into the safety loop.

The injected errors might be divided into different categories such as:

- CANopen Safety communication errors such as: exceeding the refresh time (SCT) or exceeding the re-transmit time (SVRT)
- CAN hardware errors such as: CAN-H wire connected to voltage supply or ground, CAN-L wire connected to voltage supply or ground, etc.
- *Sensor*1 or *Sensor*2 failure such as: ON or OFF failure as well as simultaneous failures of the sensors
- safety relay failure such as: ON or OFF failure as well as simultaneous failures of the safety relay

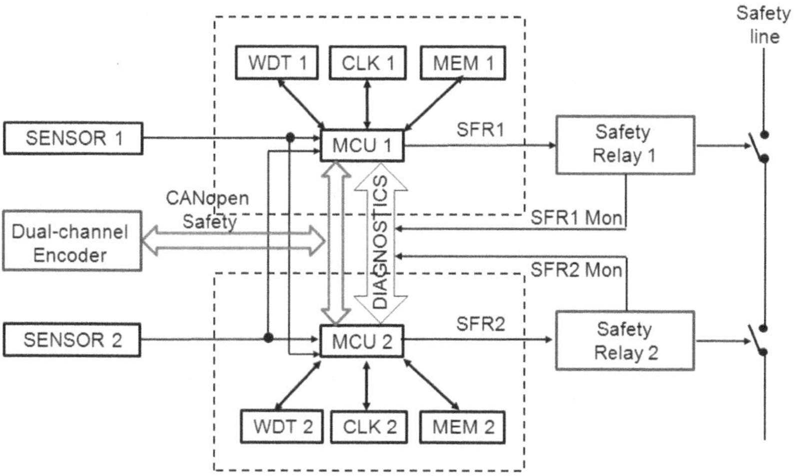

Fig. 8.23 The hardware architecture of the safety system.

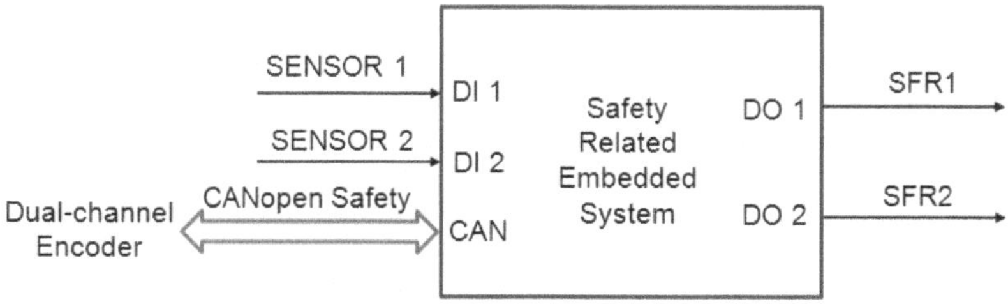

Fig. 8.24 Safety system inputs and outputs.

- power failure of the safety loop
- overvoltage or undervoltage conditions for the safety loop
- internal failures of the embedded safety loop such as: watchdog failure, microcontroller clock frequency drift, memory failure.

These are some of the most important injected error categories, which we should consider during testing the safety loop assigned to the presented linear positioning system.

It is strongly recommended to automate the tests in the HIL environment and also to consider different combinations of the above mentioned error categories.

A very simplified HIL environment (block diagram) showing some of the error injections points is presented in Fig. 8.25.

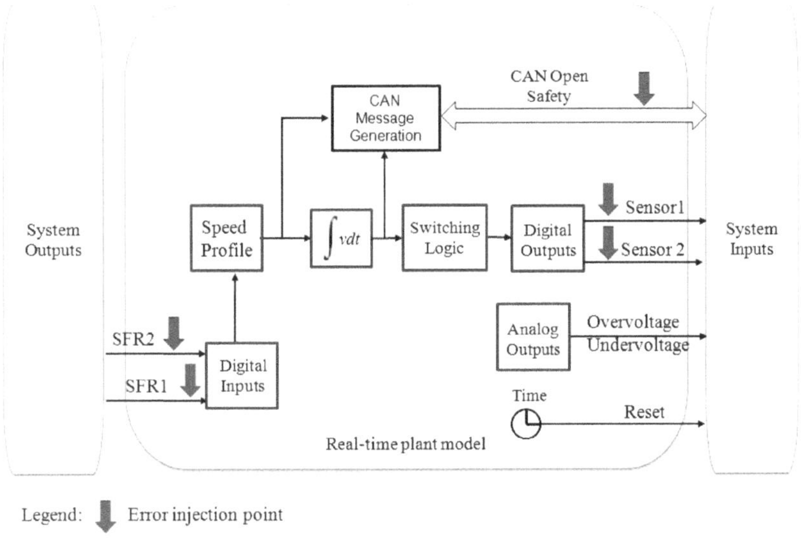

Fig. 8.25 Simplified HIL testing environment.

8.5 Memory Testing

This section is dedicated to memory testing. Memory is essential in all embedded products, it is hard to find an embedded system that does not include a certain type of memory. Furthermore, density and circuit complexity of memory chips are continuously increasing, so testing becomes more important.

A good memory testing procedure is a trade-off between the time required to apply the test and the degree of fault coverage obtained. Memory chips are tested: on the manufacturing line, during start-up a memory check is usually performed and in case of safety related embedded systems, in background a memory testing procedure is very frequently running.

Memories are divided in two big categories:

- volatile memories, also known as random access memories (RAMs), which may be static RAMs (SRAMs) or dynamic RAMs (DRAMs)
- non-volatile memories are read-only memories (ROMs), electrically erasable memories (EEPROMs), UV erasable memories (UVEPROMs) and flash memories (used whenever high-density non-volatile memory is needed).

8.5.1 RAM Fault Models

In this section we focus on RAM fault models and testing methods [1, 60]. The functional model of a RAM is shown in Fig. 8.26. The storage elements are usually organized to form an array, i.e. an array consist of r memory words having $c - bit$ length, in total $n = rxc$ memory cells. The address from the Memory Address Register (MAR) is decoded to select the appropriate cell to be accessed. Writing and reading is performed using the Write Driver and Sense Amplifiers. The data to be written to the memory or read from the memory is usually placed into a Memory Data Register (MDR). In case of cell oriented memory, one and only one memory cell is affected by the read or write operation at a time. In case of word-oriented memory, all the bits in the word are written or read simultaneously.

RAM integrated circuits are subject to the same defects as logic ICs. The defects include the presence and absence of material due to imperfection in the manufacturing process. In addition DRAMs, due to long sensing wires are more susceptible than other ICs to electromagnetic inferences and noise. Memory cells are subject to single-event upsets (SEUs), which are induced by charged particles penetrating an electronic circuit.

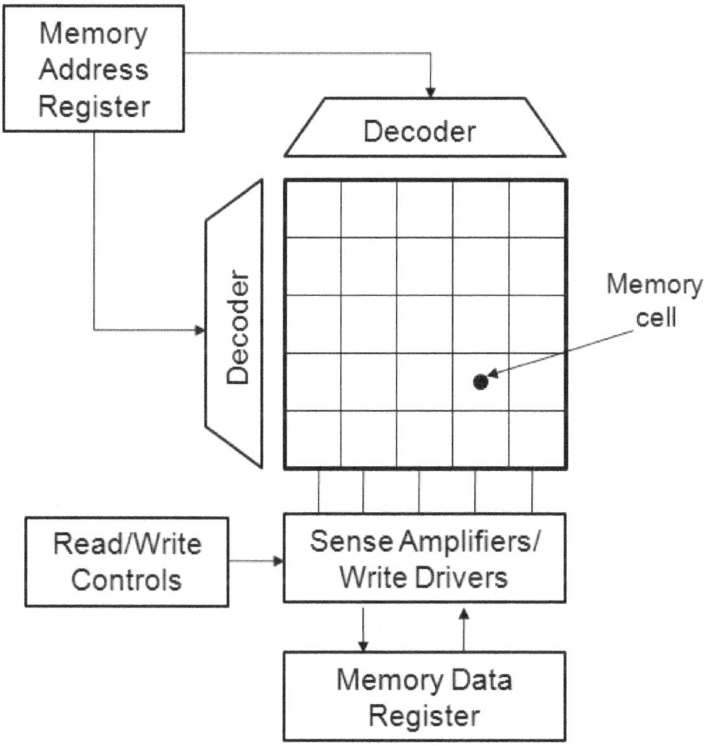

Fig. 8.26 RAM functional model.

A correctly functioning RAM shall be able to: store 1 and 0 in every cell, to make the transition to 1 to 0 and vice versa, each cell returns to its value after reading. In addition, the cells are expected to retain the information.

A functional test that would be capable of detecting all faults in these operations, would be possible, but not practical due to its time length.

Different type of memory faults are defined as follows:

- stuck-at faults are defined as functional faults that make the cells behave as if they have a 1 or a 0 stored permanently in them
- transition faults are a special case of stuck-at faults. A transition fault occurs, when a cell cannot make the transition 0 to 1 or 1 to 0. To detect such a fault it is important to make every cell undergo 0 to 1 and 1 to 0 transitions and to check the results.
- coupling faults are defined as an unintended change in one cell due to an intended change in another cell. There are three types of coupling faults:
  - inversion coupling occurs when one cell's transition causes the inversion of another cells value
  - idempotent fault occurs when one cells transition causes a particular (logic value 0 or 1) transition in the other cell
  - bridging and state coupling faults are special type of coupling faults. They occur when a certain state in one cell causes another specific state in another cell. They are caused by logic coupling rather than by transition. A bridging fault occurs when two or more lines are shorted together.
- pattern-sensitive faults occur when due to different activities of some cells an unintended transition occurs in another cell. These faults are caused mainly by the undesired interferences between cells due to the high packed density of the cells.

Related to pattern-sensitive faults we remark, that practically is not possible to examine all possible effects from some cells activities to all other cells, since the test will require very long time. It is more practical to examine only the cells in the neighborhood. These are called neighborhood pattern sensitive faults (NPSF). However, NPSF are not very common and require very high complexity tests, thus pattern-sensitive faults are rarely tested in case of industrial embedded system.

There are several testing algorithms for RAMs. An effective test is a test that detects as many faults as possible with the fewer possible patterns. The length of the tests is measured in terms of the size n of the memory array. The memory is arranged in a rxc array, where c is the word length, then $n = rxc$.

8.5.2 Testing for Stuck-at Faults

In this section different testing algorithms intended to detect stuck-at faults in semiconductor RAMs are described

The MSCAN Test Procedure

A trivial test procedure called MSCAN is illustrated below. Assume that the memory contains n words, a word with address A_i is denoted by W_i.

A write operation is denoted as:
Write: $W_i \leftarrow \alpha$ where α is the all 0 word or the all 1 word.
In a similar way the read operation:
Read: $W_i(=\alpha)$ where the value in the parenthesis is the expected value of W_i.

The MSCAN test algorithm

For i = 0,..., n-1 do:
Write: $W_i \leftarrow 0$
Read: $W_i(=0)$
Write: $W_i \leftarrow 1$
Read: $W_i(=1)$

This test writes each word, first with a 0 and then with a 1, and verifies the writing with two read statements. The above test procedure will not detect any stuck-at fault in the memory address register or in the decoder. The complexity of the MSCAN test is $4n$.

The ATS Test Procedure

Knaizuk and Hartmann developed a test algorithm called ATS (algorithmic test sequence), which detects any combination of stuck-at-0 or stuck-at-1 multiple faults in RAM [1]. Assume that the memory contains n words, each denoted as W_i for $0 \leq i \leq n-1$. The memory words are grouped into three partitions, G_0, G_1 and G_2, such that:

$G_0 = \{W_i | i = 0(modulo3)\};$
$G_1 = \{W_i | i = 1(modulo3)\};$
$G_2 = \{W_i | i = 2(modulo3)\};$

The ATS test algorithm is described below and is shown in Fig. 8.27.

The ATS test algorithm

1. For all $W_i \in G_1 \cup G_2$
   Write: $W_i \leftarrow 0$

2. For all $W_i \in G_0$
   Write: $W_i \leftarrow 1$
3. For all $W_i \in G_1$
   Read: $W_i \leftarrow 0$
4. For all $W_i \in G_1$
   Write: $W_i \leftarrow 1$
5. For all $W_i \in G_2$
   Read: $W_i = 0$
6. For all $W_i \in G_0 \cup G_1$
   Read: $W_i (= 1)$
7. For all $W_i \in G_0$
   Write: $W_i \leftarrow 0$
   Read: $W_i (= 0)$
8. For all $W_i \in G_2$
   Write: $W_i \leftarrow 1$
   Read: $W_i (= 1)$

Partition \ Step	1	2	3	4	5	6	7	8
G_0		Wr 1				R 1	Wr 0, R 0	
G_1	Wr 0		R 0	Wr 1		R 1		
G_2	Wr 0				R 0			Wr 1, R 1

Fig. 8.27 ATS algorithm (Wr means write and R means read).

Nair developed a test sequence called Modified ATS (MATS) as well as the MATS+ test sequence that detects any combination of stuck-at faults in a RAM, irrelevant of the design of the decoder combinational circuit. It is described below and is shown in Fig. 8.28. The MATS+ test sequence complexity is $5n - 2$.

The MATS+ test algorithm

Sequence 1: Address: $A_1, ..., A_{n-1}$ $\{\Uparrow (w0)\}$
Sequence 2: Address: A_0 $\{(w1)\}$
 Address: $A_1, ..., A_{n-1}$ $\{\Uparrow (r0, w1)\}$
Sequence 3: Address: $A_0, ..., A_{n-2}$ $\{\Uparrow (r1, w0)\}$
 Address: A_{n-1} $\{(r1)\}$
Sequence 4: Address: $A_0, ..., A_{n-2}$ $\{\Downarrow (r0)\}$

where the $\Uparrow$ and $\Downarrow$ symbols show up addressing and down addressing in the memory.

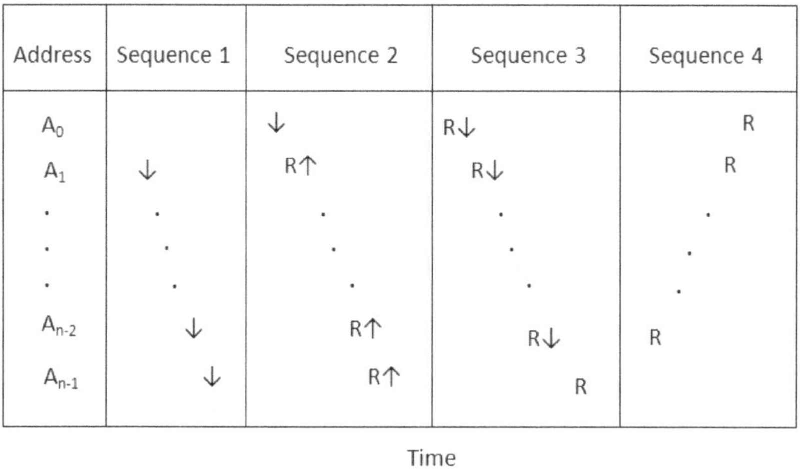

Fig. 8.28 MATS+ algorithm.

8.5.3 Testing for Coupling Faults

An important type of fault that can cause a semiconductor RAMs to function incorrectly is the cell's coupling fault. In this section we discuss several procedures that adopt the coupling fault model.

Marching 1's and 0's Procedure

Assume that the RAM consists of n cells. The Marching 1's and 0's test procedure is given below and is shown in Fig. 8.29.

The MARCH test algorithm

Step 1: Write: $C_1 \leftarrow 0$ for all cells
Step 2: For i = 0,...,n-1 do:
  Read: $C_i (= 0)$
  Write: $C_i \leftarrow 1$
  Read: $C_i (= 1)$
Step 3: For i = n-1,..., 1 do:
  Read: $C_i (= 1)$
  Write: $C_i \leftarrow 0$
  Read: $C_i (= 0)$
Step 4: Repeat Steps 1-3, interchanging 0s and 1s.

It is interesting to note that all the components of the MATS procedure exists in the MARCH test. Thus, the MARCH test covers all stuck-at faults. The disadvantage of the MARCH test is that does not detect some coupling faults such as:

Cell Nr.	Sequence 1	Sequence 2	Sequence 3	Sequence 4	Sequence 5	Sequence 6
1	0	R↑R	R↓R	1	R↓R	R↑R
2	0	R↑R	R↓R	1	R↓R	R↑R
.	.	.	.	.	.	.
.	.	.	.	.	.	.
.	.	.	.	.	.	.
n-1	0	R↑R	R↓R	1	R↓R	R↑R
n	0	R↑R	R↓R	1	R↓R	R↑R

Fig. 8.29 MARCH 1's and 0's algorithm.

- the coupling fault between cells i and j for $j > i$
- any two coupling faults that mask each other.

GALPAT Test Procedure

GALPAT or Galloping 1's and 0's is a very famous testing algorithm used by the industry to test RAMs. The GALPAT procedure is as follows:

The GALPAT test algorithm

Step 1: Write: $C_1 \leftarrow 0$ for all cells
Step 2: For i = 0,...,n-1 do:
  Read: $C_i (= 0)$
  Write: $C_i \leftarrow 1$
   For all $j \neq i$
    Read: $C_j (= 0)$ (test that no cell is disturbed)
    Read: $C_i (= 1)$ (check that test bit is still correct)
  Write: $C_i \leftarrow 0$
Step 4: Repeat Steps 1 and 2, interchanging 1s and 0s.

A major disadvantage of this algorithm is its length $O(n^2)$, which makes it impractical for large memories. This test verifies that every cell can be set to both 0 and 1 without causing any other cell to change its state, so it proves that there is no coupling between any two cells. Also, all stuck-at faults in the memory and in the decoder are covered by the test. The complexity of the GALPAT test is $4n^2 + 2n$. The GALPAT procedure is considered to be inefficient because its fault coverage does not justify its length in time.

Nair, Thatte and Abraham's (NTA) Test Algorithm

The necessary and sufficient conditions for all faults in the memory cell array to be detected by a test sequence according to [60] are mentioned bellow:

Condition 1: each cell must undergo a 0-to-1 and 1-to-0 transition and must be read after each transition and before undergoing any subsequent forced transition. If this condition is not satisfied a stuck-at fault will not be detected.

Condition 2: for every pair of cells (i, j) cell i must be read after cell j makes a transition and before cell i and j make any further transitions, for all possible states (two states) of cell i, and all possible transitions in cell j (two transitions). If this is not satisfied, a coupling fault between a pair of cells may not be detected.

Condition 3: for every cell triple (i, j, k), if the test makes a transition in cell j from y to $\overline{y}$ after cell i makes a transition from x to $\overline{x}$ and before cell k in state z is read, then the test must possess another sequence, where either

- cell k in state z is read after an x to $\overline{x}$ transition in cell i and before a y to $\overline{y}$ transition in cell j, or
- cell k in state $\overline{z}$ is read after a y to $\overline{y}$ transition in cell j and before a x to $\overline{x}$ transition in cell i.

with $[x, y, x \in \{0, 1\}]$.

If this condition is not satisfied, a coupling fault between cell j and cell k may mask the effect of another coupling fault between cell i and cell k, and prevent the fault from being detected.

The number of operations required to test the entire memory is about $30n$, where n is the number of cells in the memory. This procedure is very efficient compared with GALPAT test because of its high fault coverage and its low execution time.

The Nair, Thatte and Abraham's test procedure

Initialize:	$\{\Uparrow (w0)\}$
Sequence 1:	$\{\Uparrow (r0, w1); \Downarrow (r1)\}$
Sequence 2:	$\{\Uparrow (r1, w0); \Downarrow (r0)\}$
Sequence 3:	$\{\Downarrow (r0, w1); \Uparrow (r1)\}$
Sequence 4:	$\{\Downarrow (r1, w0); \Uparrow (r0)\}$
Sequence 5:	$\{\Uparrow (r0, w1, w0); \Downarrow (r0)\}$
Sequence 6:	$\{\Downarrow (r0, w1, w0); \Uparrow (r0)\}$
Reset:	$\{\Uparrow (w1)\}$
Sequence 7:	$\{\Downarrow (r1, w0, w1); \Uparrow (r1)\}$
Sequence 8:	$\{\Uparrow (r1, w0, w1); \Downarrow (r1)\}$

where, $\Uparrow$ means up addressing order 0 to address n-1 and $\Downarrow$ means down addressing order n-1 to address 0

In Table 8.6 the complexity and the fault coverage of different test procedures are summarized. According to IEC 61508 the diagnostic coverage of the GALPAT test is 99%, so the NTA test procedure, which has a similar fault coverage, it can be considered to have 99% diagnostic coverage.

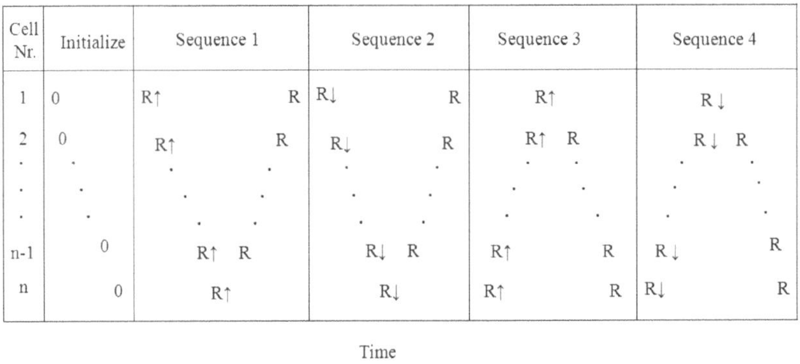

Fig. 8.30 Nair, Thatte and Abraham's test sequences from 1 to 4.

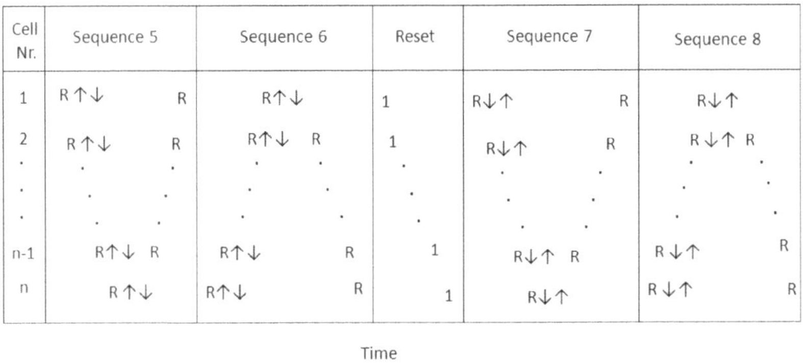

Fig. 8.31 Nair, Thatte and Abraham's test sequences from 5 to 8.

Depending on the testing approach - word based or cell based - in Table 8.6 the test complexity is expressed in terms of memory words or memory cells. For example the test complexity of the MSCAN algorithm is $4n$, where n is the number of memory words going to be tested. The test complexity of the MARCH algorithm is $14n$, where n is the number of memory cells to be tested.

Assuming a $10ns$ time for read or a write operation for RAM, in Table 8.7 we compared the test execution time of modified GALPAT test as well as the NTA test algorithm. The GALPAT test takes prohibitively long time for large memories, on the other hand the NTA test algorithm is a viable alternative for testing large memories.

As a general remark, for embedded safety systems, where the memory testing algorithm is running as a background task, the implementation requires special attention, since might lead to performance degradation or undesirable jitter effect.

Table 8.6 Complexity and fault coverage of different test procedures

Method	Approach	Complexity	Fault coverage: Stuck-at faults	Fault coverage: Coupling faults
MSCAN	word based	$4n$	Covers memory array faults, but does not cover decoder or MAR faults.	No
ATS	word based	$4n$	Covers all, stuck-at faults for RAMs with wired-OR logic behavior and noncreative decoder design.	No
MATS+	word based	$5n-2$	Covers all faults, for RAM with arbitrary wired logic behavior and arbitrary decoder design.	No
MARCH	cell based	$14n$	All	Does not cover all coupling faults.
Modified GALPAT	cell based	$4n^2+4n$	All	All
NTA	cell based	$30n$	All	All

Table 8.7 Test execution time [s]

Method	Memory size [bit]: 1K x 1	1K x 8	1M x 1	1M x 8
Modified GALPAT: Complexity: $4n^2+4n$	10 s	655 s	$4 \cdot 10^4$ s	$2.56 \cdot 10^6$ s
Nair, Thatte and Abraham test method: Complexity: $30n$	$4.8 \cdot 10^{-3}$ s	$38.4 \cdot 10^{-3}$ s	0.3 s	2.4 s

References

[1] Abadir, M.S., Reghbati, K.H.: Functional Testing of Semiconductor Random Access Memories. Computing Surveys 15(3), 175–198 (1983)

[2] Adachi, S.: System Identification - Theory for the User. The Society of Instrument and Control Engineers, Japan (1993) (in Japanese)

[3] Adachi, S., Ogawa, T., Konno, R.: A System Identification Method for Linear Regression Models Based on Support Vector Machine. Trans. of the Society of Instrument and Control Engineers Japan 37(12), 1189–1191 (2001)

[4] Adlgostar, R., Azimian, H., Taghirad, H.D.: Robust H_∞, H_2/H_∞ Controller for Rotational/Translational Actuator (RTAC). In: Proc. of IEEE Conf. on Control Applications, pp. 705–710 (2006)

[5] Ahmad, S.S., Lew, J.-S., Keel, L.H.: Robust Control of Flexible Structures against Structural Damage. IEEE Trans. on Control Systems Technology 8(1), 170–181 (2000)

[6] Ando, N., Szemes, P.T., Korondi, P., Hashimoto, H.: Friction compensation for 6DOF Cartesian coordinate haptic interface. In: IEEE/RSJ Int. Conf. on Intelligent Robots and System, vol. 3, pp. 2893–2898 (2002)

[7] Araki, M., Taguchi, H.: Two-degree-of-freedom PID controllers. Int. Journal of Control, Automation, and Systems 1(4), 401–411 (2003)

[8] Åström, K.J., Wittenmark, B.: Adaptive Control, 2nd edn. Addison-Wesley (1995)

[9] Balas, G.J., Doyle, J.C.: Identification of Flexible Structures for Robust Control. IEEE Control Systems Magazine, 51–58 (June 1990)

[10] Balas, G.J.: Synthesis of Controllers for the Active Mass Driver System in the Presence of Uncertainty. Earthquake Engng. Struct. Dyn. 27, 1189–1202 (1998)

[11] Bernstein, D.S., Michel, A.N.: A Chronological Bibliography on Saturating Actuators. Int. Journal of Robust and Nonlinear Control, Special Issue in Saturating Actuators 5, 375–380 (1995)

[12] Bernstein, D.S.: What makes some control problem hard? Control Systems Magazine 22(4), 8–19 (2002)

[13] Bittar, A., Sales, R.M.: $\mathcal{H}_2$ and $\mathcal{H}_\infty$ control for MagLev vehicles. IEEE Control Systems 18(4), 18–25 (1998)

[14] Boulet, B., Francis, B.A., Hughes, P.C., Hong, T.: Uncertainty Modeling and Experiments in H_∞ Control of Large Flexible Space Structures. IEEE Trans. on Control Systems Technology 5(5), 504–519 (1997)

[15] Breneman, S.E., Allison Smith, H.: Design of $\mathcal{H}_\infty$ Output Feedback Controllers for the AMD Benchmark Problem. Earthquake Engng. Struct. Dyn. 27, 1277–1289 (1998)
[16] Briz, F., Degner, M.W., Garcia, P., Lorenz, R.D.: Comparison of saliency-based sensorless control techniques for AC machines. IEEE Trans. on Industry Applications 40(4), 1107–1115 (2004)
[17] Casella, F., Locatelli, A., Schiavoni, N.: Modeling and Control for Vibration Suppression in Large Flexible Structure with Jet Thrusters and Piezoactuators. IEEE Trans. on Control Systems Technology 10(4), 589–599 (2002)
[18] Chiang, R.Y., Safonov, M.G.: Robust Control Toolbox, User's Guide. The MathWorks, Inc. (1992)
[19] Chopra, A.K.: Dynamics of Structures - Theory and Applications of Earthquake Engineering. Prentice-Hall, New York (1995)
[20] Corriga, G., Giua, A., Usai, G.P.: An Implicit Gain-Scheduling Controller for Cranes. IEEE Trans. on Control Systems Technology 6(2), 15–20 (1998)
[21] Coustal, P., Michelin, J.M.: Industrial Application of an H-infinity Design Method for Flexible Structures. IEEE Control Systems Magazine, 49–54 (August 1994)
[22] DeRusso, P.M., Roy, R.J., Close, C.M., Desrochers, A.A.: State Variables for Engineers. John Wiley & Sons Inc., New York (1992)
[23] Dorf, R.C., Bishop, R.H.: Modern Control Systems, 9th edn. Prentice Hall, New Jersey (2000)
[24] Douma, S.G., van den Hof, P.M.J.: Relations between uncertainty structures in identification for robust control. Automatica 41, 439–457 (2005)
[25] Doyle, J.C., Francis, B.A., Tannenbaum, A.R.: Feedback Control Theory. Mcmillan, New York (1992)
[26] Doyle, J.C., Glover, K., Khargonekar, P.P., Francis, B.: State-space solutions to standard H_2 and H_∞ control problems. IEEE Trans. on Automatic Control 34(8), 831–847 (1989)
[27] Dullerud, G.E., Paganini, F.: A Course in Robust Control Theory - Convex Approach. Springer, New York (2000)
[28] Forrai, A., Hashimoto, S., Isojima, A., Funato, H., Kamiyama, K.: Gray box identification of flexible structures: application to robust active vibration suppression control. Journal of Earthquake Engineering and Structural Dynamics 30, 1203–1220 (2001)
[29] Forrai, A., Hashimoto, S., Funato, H., Kamiyama, K.: Structural Control Technology - System Identification and Control of Flexible Structures. IEE Computing & Control Engineering Journal 12(6), 257–262 (2001)
[30] Forrai, A., Tanoi, T., Hashimoto, S., Funato, H., Kamiyama, K.: Robust Controller Design with Hard Constraints on the Control Signal. Journal of Electrical Engineering, Archiv fuer Elektrotechnik 83(4), 179–186 (2001)
[31] Forrai, A., Funato, H., Yanagita, Y., Kato, Y.: Fuel Cell Parameter Estimation and Diagnostics. IEEE Trans. on Energy Conversion 20(3), 668–675 (2005)
[32] Forrai, A., Ueda, T., Yumura, T.: Electromagnetic Actuator Control: a Linear Parameter-Varying (LPV) Approach. IEEE Trans. on Industrial Electronics 54(3), 1430–1441 (2007)
[33] Furuya, T., Toyoda, Y., Hori, Y.: Implementation of advanced adhesion control for electric vehicle. In: Proc. of Advanced Motion Control, vol. 2, pp. 430–435 (1996)

[34] Ganssle, J.: The art of designing embedded systems, 2nd edn. Elsevier Inc. (2008)

[35] Gahinet, P., Nemirovski, A., Laub, A.J., Chilali, M.: LMI Control Toolbox, User's Guide. The MathWorks, Inc. (1994)

[36] Goodwin, G.C., Graebe, S.F., Salgado, M.E.: Control System Design. Prentice Hall, Upper Saddle River (2001)

[37] Glover, K.: Model Reduction: A tutorial on Hankel-norm methods and lower bounds on L^2 errors. In: Proc. of the 10th IFAC World Congress, pp. 9–22 (1987)

[38] Haber, R., Keviczky, L.: Identification of linear systems having signal-dependent parameters. Int. J. Systems Sci. 16(7), 869–884 (1985)

[39] Halim, D., Moheimani, S.O.R.: Spatial H_2 Control of a Piezoelectric Laminate Beam: Experimental Implementation. IEEE Trans. on Control Systems Technology 10(4), 533–546 (2002)

[40] Hamamci, S.E.: A robust polynomial-based control for stable process with time delay. Journal of Electrical Engineering, Archiv fuer Elektrotechnik 87(3), 163–172 (2005)

[41] Hashimoto, S., Funato, H., Hara, K., Kamiyama, K.: DVFC-Based Identification and Its Application to Robust Vibration Suppression Control of Flexible Structures. Trans. IEEJ. D-119(6), 888–894 (1999)

[42] Horowitz, I.M.: Synthesis of Feedback Systems. Academic Press, New York (1963)

[43] Hyunbae, K., Harke, M.C., Lorenz, R.D.: Sensorless control of interior permanent-magnet machine drives with zero-phase lag position estimation. IEEE Trans. on Industry Applications 39(6), 1726–1733 (2003)

[44] Ionescu, V., Varga, A.: Control theory - robust synthesis and numerical methods. All Publishing House (1994) (in Romanian)

[45] Isermann, R., Balle, P.: Trends in the application of model-based fault detection and diagnosis of technical processes. Control Engineering Practice 5(5), 709–719 (1997)

[46] Khalil, H.K.: Nonlinear Systems, 3rd edn. Prentice Hall (2002)

[47] Lindlau, J.D., Knospe, C.R.: Feedback linearization of an active magnetic bearing with voltage control. IEEE Trans. on Control Systems Technology 10(1), 21–31 (2002)

[48] Kwakernaak, H.: Robust control and H_∞-optimization - tutorial paper. Automatica 29(2), 255–273 (1993)

[49] Landau, I.D., Karimi, A.: A recursive algorithm for ARMAX model identification in closed loop. IEEE Trans. on Automatic Control 44(4), 840–843 (1999)

[50] Leith, D.J., Leithead, W.E.: Survey of gain-scheduling analysis and design. Int. Journal of Control 73(11), 1001–1025 (2000)

[51] Ljung, L.: System Identification - Theory for the User. Prentice-Hall, New York (1987)

[52] Ljung, L.: Systems Identification Toolbox, User's Guide. The MathWorks, Inc. (1997)

[53] Lenz, K., Özbay, H., Tannenbaum, A., Turi, J., Morton, B.: Robust control design for flexible beam using a distributed-parameter H_∞-method. In: Proc. of 28th Conference on Decision and Control, pp. 2673–2678 (1989)

[54] Lo, K., Kimura, H.: Recursive estimation methods for discrete systems. IEEE Trans. on Automatic Control 48(11), 2019–2024 (2003)

[55] Ma, L., Feng, X., Peng, Z.: Integrated Design and Implementation of Embedded Control Systems with Scilab. Sensor 8, 5501–5515 (2008)
[56] Manabe, S.: Analytical weight selection for LQ design. Proceedings of the 8th Workshop on Astrodynamics and Flight Mechanics, Sagamihara, ISAS (1998)
[57] Manabe, S.: A low cost inverted pendulum system for control system education. In: Proceedings of the 3rd IFAC Symposium on Advances in Control Education, Tokyo (1994)
[58] McFarlane, D.C., Glover, K.: Robust Controller Design Using Normalized Coprime Factor Plant Description. LNCIS. Springer, Berlin (1989)
[59] Menhold, P.H., Pearson, R.K., Allgöwer, F.: Online outlier detection and removal. In: Proc. of the 7th Mediterranean Conference on Control and Automation, Haifa, Israel, pp. 1110–1133 (June 1999)
[60] Nair, R., Thatte, S.M., Abraham, J.A.: Efficient algorithms for testing semiconductor random-access memories. IEEE Trans. on Computers C-27(6), 572–576 (1978)
[61] Özbay, H.: Introduction to Feedback Control Theory. CRC Press (1999)
[62] Reinelt, W.: H_∞ Loop Shaping for Systems with Hard Bounds. In: Proc. on Quantitative Feedback Theory and Robust Frequency Domain Methods, Durban, South Africa, pp. 89–103 (1999)
[63] Reinelt, W.: Loop Shaping of Multivariable Systems with Hard Constraints on the Control Signal. Electrical Engineering, Archiv fuer Elektrotechnik 83, 169–177 (2001)
[64] Rohrs, C.E., Melsa, J.L., Schultz, D.G.: Linear Control Systems. McGraw-Hill, New York (1993)
[65] Rojo-Alvarez, J.L., Martinez-Ramon, M., Prado-Cumplido, M., Artes-Rodriquez, A., Figueiras-Vidal, A.R.: Support Vector Method for Robust ARMA System Identification. IEEE Trans. on Signal Processing 52(1), 155–164 (2004)
[66] Rugh, W.J., Shamma, J.S.: Research on gain scheduling - survey paper. Automatica 36, 1401–1425 (2000)
[67] Saberi, A., Stoorvogel, A.A., Sannuti, P.: Control of Linear Systems with Regulation and Input Constraints. Springer, Berlin (2000)
[68] Saberi, A., Han, J., Stoorvogel, A.A.: Constrained Stabilization Problems for Nonlinear Plants. In: Proc. of the American Control Conference, Chicago, pp. 4393–4397 (2000)
[69] van der Schaft, A.: L_2-Gain and Passivity Techniques in Nonlinear Control. Springer-Verlag, London Limited (2000)
[70] Shamma, J.S., Athans, M.M.: Guaranteed Properties of Gain Scheduled Control for Linear Parameter-Varying Plants. Automatica 27(3), 559–564 (1991)
[71] Spencer Jr., B.F., Sain, M.K.: Controlling Buildings: A New Frontier in Feedback. IEEE Control Systems Magazine on Emerging Technology 17, 19–35 (1997)
[72] Spencer Jr., B.F., Dyke, S.J., Deoskar, H.S.: Benchmark problems in structural control: Part I - Active Mass Driver System. Journal of Earthquake Engineering and Structural Dynamics 27, 1127–1139 (1998)
[73] Stein, G.: Respect the unstable. IEEE Control Systems Magazine, 12–25 (August 2003)

[74] Takanashi, H., Adachi, S., Wakui, S.: Independent modal control system design method via modal matrix estimation based on subspace method. In: The 6th International Conference on Motion and Vibration Control, Saitama, Japan (August 2002)

[75] Thompson, S.K.: Sample size for estimating multinominal proportions. The American Statiscian 41(1), 42–46 (1987)

[76] Vapnik, V.N.: Statistical Learning Theory. John Willey & Sons (1998)

[77] Vapnik, V.N.: An Overview of Statistical Learning Theory. IEEE Trans. on Neural Networks 10(5), 988–999 (1999)

[78] Venini, P., Mimmi, G., Tanzi, E.: Robust Vibration Control: Acceleration Feedback and μ Synthesis. In: Proc. Int. Conf. on Motion and Vibration Control, MOVIC 1996, pp. 274–278 (1996)

[79] Watakabe, M., Chiba, O., Kamada, T.: Vibration control of a large-scale test building using hybrid mass damper system - applied robust control theory. In: Proc. of 13th World Conference on Earthquake Engineering, paper No. 2510 (2004)

[80] Whorton, M.S., Calise, A.J., Hsu, C.C.: A Study of Fixed-Order Mixed Norm Designs for a Benchmark Problem in Structural Control. Earthquake Engng. Struct. Dyn. 27, 1315–1330 (1998)

[81] Xue, D., Chen, Y.Q., Atherton, D.P.: Linear Feedback Control. Society for Industrial and Applied Mathematics (2007)

[82] Young, P.M., Bienkiewicz, B.: Robust Controller Design for the Active Mass Benchmark Problem. Earthquake Engng. Struct. Dyn. 27, 1149–1164 (1998)

[83] Zhou, K., Doyle, J.C., Glover, K.: Robust and Optimal Control. Prentice-Hall, Inc., New York (1996)

[84] Functional safety of electrical/ electronic/ programmable electronic safety-related systems (IEC61508), 2nd edn. (2010)

[85] Hardware-Software Design according to IEC61508 (2008)

[86] Control Area Network (CAN) basics. Microchip Technology Inc. (1999)

[87] ISO 11898 Control Area Network (CAN) serial communication

[88] OSEK/VDX: Operating Systems Specifications 2.2.3. OSEK organization, pp. 18–36

[89] LTC 2911 Precision Triple Supply Monitor with Power-Fail Comparator. Catalog Linear Technology

[90] TPS3813J25 Processor Supervisory Circuits with Window-Watchdog. Catalog Texas Instruments

The manufacturer's authorised representative in the EU is Springer Nature Customer Service Centre GmbH, Europaplatz 3, 69115 Heidelberg, Germany. If you have any concerns regarding our products, please contact ProductSafety@springernature.com

Printed and bound by CPI Group (UK) Ltd, Croydon, CR0 4YY
01/10/2026
02225968-0008